Contents

Reflect and review 79

NEBS

MANAGEMENT

DEVELOPMENT

SUPER SERIES

THIRD EDITION

Managing Activities

Managing Lawfully
– Health, Safety and Environment

Published for

&& NEBS Management *by*

**Pergamon
Flexible
Learning**

Pergamon Flexible Learning
An imprint of Butterworth-Heinemann
Linacre House, Jordan Hill, Oxford OX2 8DP
225 Wildwood Avenue, Woburn, MA 01801-2041
A division of Reed Educational and Professional Publishing Ltd

A member of the Reed Elsevier plc group

OXFORD   AUCKLAND   BOSTON
JOHANNESBURG   MELBOURNE   NEW DELHI

First published 1986
Second edition 1991
Third edition 1997
Reprinted 1998, 1999, 2000

© NEBS Management 1986, 1991, 1997

British Library Cataloguing in Publication Data
A catalogue record for this book is available from the British Library

ISBN 0 7506 3301 8

FOR EVERY TITLE THAT WE PUBLISH, BUTTERWORTH-HEINEMANN
WILL PAY FOR BTCV TO PLANT AND CARE FOR A TREE.

Whilst every effort has been made to contact copyright
holders, the author would like to hear from anyone
whose copyright has unwittingly been infringed.

The views expressed in this work are those
of the authors and do not necessarily reflect
those of the National Examining Board for
Supervision and Management or of the publisher.

NEBS Management Project Manager: Diana Thomas
Author: Joe Johnson
Editor: Fiona Carey
Series Editor: Diana Thomas
Based on previous material by: Joe Johnson
Composition by Genesis Typesetting, Rochester, Kent
Printed and bound in Great Britain

Workbook introduction

Here are the workbook titles in each module which link with *Managing Lawfully – Health, Safety and Environment*, should you wish to extend your study to other Super Series workbooks. There is a brief description of each workbook in the User Guide.

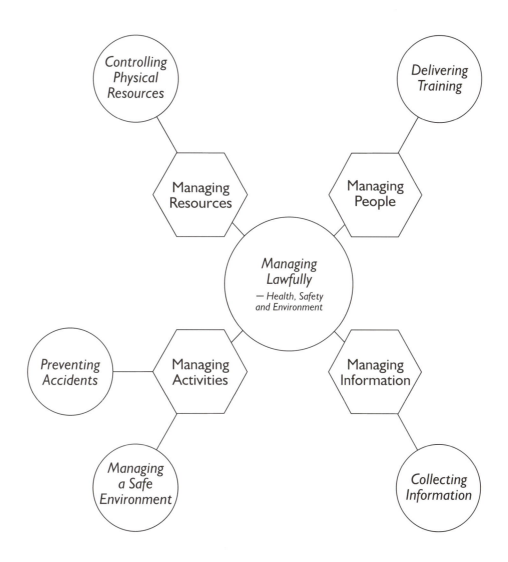

2 S/NVQ links

This workbook relates to the following element:

A1.2   Maintain healthy, safe and productive work conditions

It is designed to help you to demonstrate the following Personal Competences:

■ building teams;
■ focusing on results;
■ thinking and taking decisions;
■ striving for excellence.

3 Workbook objectives

All managers need to know enough to ensure that the work activities they control remain within the requirements of the law. In the areas of health, safety and the environment, this is becoming increasingly difficult, because so many changes in the law have been made in recent times. Unfortunately, ignorance of the law is no defence, so it's of no avail to plead 'Nobody told me!'

As a front line manager, you should make it your job to learn as much about the law as you can, even if only to help you in planning your team's work.

Also, as a team leader, you have special responsibilities for the health and safety of your team members, as well as your own. Another good reason for studying the law on health and safety is that it provides guidance on minimum standards.

If you need one further reason for reading about the law, it is this. Should you break the law, **there is a real possibility that action could be taken against you, personally, as well as against your organization.** This is especially likely to happen if a serious accident occurs as a result of your actions, or because of your failure to act.

This workbook is divided into three sessions. Sessions A and B are devoted to health and safety aspects of the law, and Session C is concerned with the environment.

In Session A you will be able to read about the background to the law on health and safety. Session B goes on to describe the principal Acts and Regulations.

Session C explains environmental law in terms of its sources, its history and the way that it is enforced, and summarizes the main statutes related to the environment.

Notes on studying this workbook.

This book contains quite a lot of detail about health and safety and environmental legislation. **You are not expected to remember it all**. The best way to tackle the workbook is to read it through, completing the Activities, and answering the self-assessment questions, in the usual way. You should be able to follow the points made, but don't feel you have to learn them all by heart.

Whenever you come across areas of law that seem particularly relevant to you and your job, make a note to remind yourself to find out more. There is a list of extensions at the back of the book, starting on page 83; alternatively, there may be people in your organization who can give you guidance.

3.1 Objectives

When you have completed this workbook you will be better able to:

■ identify the most important laws related to health and safety;
■ find out more about laws that are especially relevant to the work you do;
■ explain to your team how the law affects them, and the duties imposed by the law on everyone at work;
■ understand the law on the environment.

4 Activity planner

The following Activities require some planning so you may want to look at these now.

■ In Activity 13 you are asked to think about the way you give information about health and safety to your team at present, and how you might improve your effectiveness in this respect.
■ For Activity 23, you are expected to look at the training you give on hazardous substances in your work area.

Portfolio of evidence

Some or all of these Activities may provide the basis of evidence for your S/NVQ portfolio. All Portfolio Activities and the Work-based assignment are signposted with this icon.

The icon states the elements to which the portfolio activities and Work-based assignment relate.

In the Work-based assignment you are required to investigate how well your part of the organization complies with one of the sets of regulations we discuss.

Session A Background to health and safety legislation

1 Introduction

> 'The gross ill-usage of little boys as sweeps, by masters who found it cheaper to drive them through soot-choked chimneys than to use a long brush, had long been exposed to the public indignation, but in vain. In 1875 Shaftesbury ... obtained the passing of an Act that at last cured the evil.'
>
> G. M. Trevelyan, *Illustrated English Social History*[1]

The health and safety laws brought in during the latter part of the nineteenth century were designed to correct only the worst abuses. To give another example, the Consolidating Act of 1878 increased the minimum age of child labour in textile factories (but not other factories) to ten years. Even until the middle of the twentieth century, the health and safety of people at work were poorly safeguarded by today's standards.

Since that time, many acts of parliament and regulations have been introduced related to health and safety at work.

A key turning point took place in 1974, when the Health and Safety at Work etc. Act (HSWA) was passed, and was welcomed as a major step forward. HSWA applies to all work premises, and sets out clear principles and guidelines for the promotion of health and safety, for both employers and employees.

We will look at the contents of this act in this session, and discuss the different levels of duty imposed by the law.

The next section is a brief review of the functions and responsibilities of safety representatives and safety committees.

But we begin with a preface to our subject.

There are some who think that things have swung too far the other way, and that the employer is now besieged by regulations that govern every aspect of employment!

[1] Volume 4, page 157. Pelican Books, 1964. Penguin Books Ltd, 27 Wrights Lane, London W8 5TZ.

2 Introduction to health and safety legislation

We need to start by seeing health and safety in the context of the law in general.

2.1 Sources of law

Although Scottish law has continued to develop along different lines from English law since the Act of Union in 1703, and is partly derived from Scottish common law, all the acts and regulations we will discuss in this workbook are also applicable in Scotland.

English law (applicable in England and Wales) is derived from three principal sources:

■ statute law

  acts of parliament, such as the Health and Safety at Work etc. Act 1974, together with subordinate legislation (sometimes called 'statutory instruments'), such as the Management of Health and Safety at Work Regulations 1992.

■ common law

  based on case law – the decisions made in courts over the centuries. Once a judgement is made, a **precedent** is established. A court is bound to follow earlier decisions made in higher courts, or in courts at the same level.

■ contract law

  governing agreements between two parties. Contract law does not play much part in health and safety.

2.2 Civil law and criminal law

Both criminal law and civil law are important to organizations in terms of health and safety.

Anyone committing a crime has offended against the state, and is in breach of criminal law. If an organization fails to comply with its statutory health and safety duties, then it or its officers may be prosecuted under criminal law. If guilt is proved 'beyond reasonable doubt', the offender may be punished by the court by having a fine imposed. In theory at least, jail sentences can also be passed on individuals.

Under civil actions, a plaintiff sues a defendant, usually for damages, that is, financial compensation. As an example, an individual may sue an employer if he or she is injured at work. A lesser standard of proof applies in civil actions: cases have to be proved 'on the balance of probabilities', rather than 'beyond reasonable doubt'.

Activity 1

2 mins

Briefly describe the **two** main ways in which organizations may have legal actions brought as a result of an accident at work.

_____

_____

_____

_____

The following diagram shows the possible routes that could be taken through the legal system, following an accident at work.

The two main routes are through the civil and criminal courts. The third route, shown on the right of the diagram, is via an industrial tribunal.

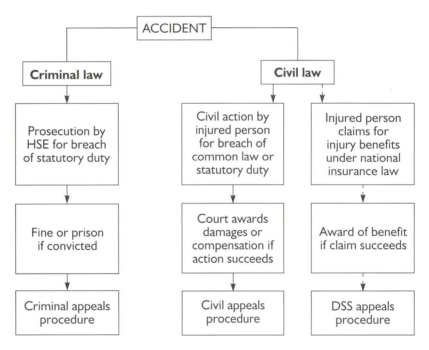

EXTENSION 1
This table is taken from page 5 of _Health and Safety Law_ by Jeremy Stranks.

Redrawn, with kind permission, from a diagram in _Health and Safety Law_, by Jeremy Stranks

2.3 The courts system

The next diagram summarizes the courts system in England and Wales.

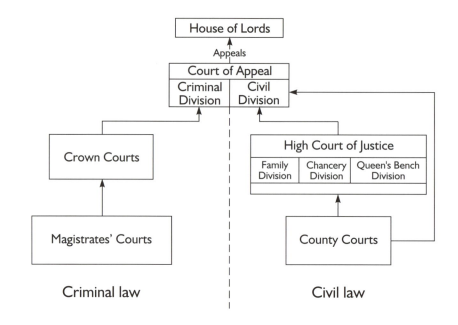

As you can see from the diagram, there are two systems, one for civil actions, and the other for criminal actions.

Magistrates' courts deal largely with the less serious criminal offences. More serious cases go to one of the Crown Courts, sometimes having been referred there by a magistrates' court. Crown Courts also hear appeals from magistrates' courts.

County courts deal only with civil matters. More complex civil cases, or ones involving large sums of money, go instead to one of the three divisions of the High Court of Justice:

- the Queen's Bench Division, which handles contract and torts;
- the Chancery Division, dealing with cases concerning land, partnerships and companies;
- the Family Division, which deals with such matters as divorce, and child custody.

Appeals, in which a higher court is asked to review the decision of a lower one, may go the Court of Appeal, then perhaps to the House of Lords. Some appeals go further than this, to the European legal system.

2.4 European law

The United Kingdom is a member state of the European Union (EU), along with, in April 1997, fourteen other nations: Austria, Belgium, Denmark, Finland, France, Germany, Greece, Holland, Ireland, Italy, Luxembourg, Portugal, Spain and Sweden.

Although we still make our own laws in the UK, membership of the EU has had a profound effect on our laws and lawmaking. One important fact is that, if ever there is any conflict, European laws take precedence over national laws in the courts.

The two principal instruments by which the European Union makes laws is through:

■ **EU regulations** that apply directly in all member countries. Actions based on EU regulations can be brought in national courts. (EU regulations should not be confused with UK regulations, many of which we will discuss in this workbook.)
■ **directives,** which bind member countries to comply with an agreed ruling. Unlike EU regulations, directives are normally made into national laws by each state. A good deal of modern health and safety legislation is the direct result of EU directives.

The **Single European Act** was made law at the beginning of 1993. Its aim is to eliminate technical barriers to trade, by introducing a new approach to technical harmonization and standards. Largely as a result of this act, national standards for health and safety within the Union are being made to conform with one another.

2.5 Approved codes of practice and guidance notes

These two kinds of documents are useful sources of information about the law.

Approved codes of practice (ACOPs) are issued by the Health and Safety Commission (HSC) as interpretations of regulations, and are intended to help people apply the law in practice. ACOPs are designed to:

■ make regulations more plain or more specific;
■ explain how regulations can be complied with, in a satisfactory way.

Example

Regulation 15 of the Workplace (Health, Safety and Welfare) Regulations 1992 states that:

1   No window, skylight or ventilator which is capable of being opened shall be likely to be opened, closed or adjusted in a manner which exposes any person performing such operation to a risk to his health or safety.

2   No window, skylight or ventilator shall be in a position when open which is likely to expose any person in the workplace to a risk to his health or safety.

Part of the ACOP for this regulation says:

153   It should be possible to reach and operate the control of openable windows, skylights and ventilators in a safe manner. Where necessary, window poles or similar equipment should be kept available, or a stable platform or other safe means of access should be provided. Controls should be so placed that people are not likely to fall through or out of the window. Where there is a danger of falling from a height devices should be provided to prevent the window opening too far.

Guidance notes may also be issued, either by the Health and Safety Commission (HSC) or the Health and Safety Executive (HSE). They include, for example, advice on action to be taken by employers in order to conform with the law.

To summarize this introduction:

■ the three sources of law are common law, statute law, and contract law;
■ it's important to distinguish between criminal law and civil law, and there is a separate court system for each; however, both are important in health and safety;
■ the UK's membership of the European Union has had a profound effect on our health and safety legislation;
■ useful documents that are intended to help people apply the law are approved codes of practice (ACOPs) and guidance notes.

3 The Health and Safety at Work etc. Act, 1974 (HSWA)

HSWA applies to all people at work, with the exception of domestic servants in a private household.

The Health and Safety at Work etc. Act is an 'enabling act'. Its purpose is to provide the legislative framework to promote, stimulate and encourage high standards of health and safety at work.

The act set up the means for gradually replacing existing health and safety laws, with regulations and approved codes of practice, as indicated in the following diagram.

HSWA

Health and Safety (First Aid) Regulations, 1981

Electricity at Work Regulations, 1989

Noise at Work Regulations, 1989

Management of Health and Safety at Work Regulations, 1992

Workplace (Health, Safety and Welfare) Regulations, 1992

and so on.

Two of the most significant clauses in HSWA read as follows:

EXTENSION 2
A Guide to the Health and Safety at Work etc. Act, 1974 is available from HSE Books.

Section 2(1): 'It shall be the duty of every employer to ensure, as far as reasonably practicable, the health, safety and welfare at work of all his employees.'

Extract from Section 7: 'It shall be the duty of every employee while at work . . . to take reasonable care for the health and safety of himself and of other persons who may be affected by his acts or omissions at work.'

Activity 2

Who in law has duties under HSWA, according to these extracts?

_____

_____

_____

Both employers and employees have duties under HSWA.

Let's look at the duties of the employer first.

3.1 The employer's overall duties under HSWA

Under HSWA, an employer has a duty:

'to ensure, as far as reasonably practicable, the health, safety and welfare at work of all his employees'.

The key words in the extract from Section 2(1) are '**as far as reasonably practicable**'. This is the 'yardstick' by which an employer's actions will be judged.

> We will discuss what 'as far as reasonably practicable' means in the next section.

To do this, the employer will need to be sure that (to give a few examples):

■ plant and equipment is safely installed, operated and maintained;
■ systems of work are checked frequently, to ensure that risks from hazards are minimized;
■ the work environment is regularly monitored to ensure that people are protected from any toxic contaminants;
■ safety equipment is inspected regularly;
■ risks to health from 'natural and artificial substances' are minimized.

HSWA also places an obligation on employers to take care of the health and safety of non-employees.

Activity 3

3
mins

Can you suggest **two** groups of people, other than employees, that an employer may have duties towards under health and safety laws?

_____

_____

_____

_____

You may have mentioned:

- self-employed people or contractors' employees working on site;
- customers who visit (for instance) shop or garage premises;
- visiting suppliers;
- other visitors;
- the general public living and working outside the worksite.

3.2 The statement of health and safety policy

HSWA further states that it is the duty of every employer of five or more people to prepare and keep up to date:

'**a written statement of his general policy** with respect to the health and safety at work of his employees, and the organization and arrangements for . . . carrying out that policy, and to bring the statement and any revision of it to the notice of all of his employees.'

This statement of general policy can be considered in three parts. Often the three parts are clearly separated.

- General policy statement

A statement of the organization's commitment to health and safety, and its obligations towards its employees. This statement should also make clear what the duties of employees are.

- Statement of organization

A statement of specific responsibilities: who is responsible for what aspects of health and safety, and how the organization is structured.

■ Statement of arrangements

A statement of the specific arrangements for dealing with health and safety matters.

There should be a clear indication of the manner in which these statements are brought to the notice of employees.

Activity 4

3 mins

What do you think is gained by making an employer provide a health and safety policy statement, as described above?

_____

_____

_____

_____

In providing a health and safety policy statement, the employer is required to think carefully about:

■ his or her responsibilities, and what must be done to meet those responsibilities;
■ how to comply with the statement to organize and implement safe systems.

Including details on specific responsibilities and on arrangements for implementing the safety policy means that employees should have a clearer idea of how to act in a safe manner and what to do if something goes wrong.

This brings us to the subject of the responsibilities of employees – we'll look at those next.

3.3 The employee's duties under HSWA

Under HSWA, employees have a duty:

■ to take reasonable care to avoid injury to themselves or to others by their work activities; and
■ to co-operate with employers and others in meeting the requirements of the law; and
■ not to interfere with or misuse anything provided to protect their health, safety and welfare.

Activity 5

Kenny works for a demolition contractor. Kenny's job involves him in having to work high up on the outside of buildings. The nature of the job means that objects fall down from where he is working, and that the conditions underfoot can be treacherous.

We know that Kenny's employer has a duty to take all practicable steps to ensure his safety. But what about Kenny himself? What steps should he take to ensure his own safety?

_____

_____

_____

_____

We can perhaps agree that Kenny should do everything possible for his own safety, including:

- wear all the protective clothing (safety helmet, boots, safety harness and so on) that is provided for him;
- check for himself that this equipment is in good condition;
- make sure he knows how to use all the equipment and machinery he has to deal with;
- communicate with his workmates so that he is aware of what hazardous conditions exist at the worksite at all times;
- behave in a sensible and careful manner, and not take unnecessary risks.

All employees have a duty to work diligently, that is, not to omit to do things that should be done to ensure safety.

Activity 6

What of the safety of others?

Jot down **three** things you would expect a member of your team to do, or **not** do, in order to help to ensure the safety of others.

_____

_____

_____

_____

Your response will be relevant to the kind of job you do. In general, you might expect a team member:

■ **to think of the safety and health of others when carrying out his or her job**

Kenny on the demolition site would be expected to warn his workmates of any hazard he is creating, such as demolishing a wall. In another kind of job, a typist in an office would be expected to make sure that cables, boxes and other obstacles are not a hazard to people walking by.

■ **to behave sensibly and responsibly in matters of health and safety**

For instance, it would be irresponsible for someone to cover up a safety notice, or to use a fire bucket for another purpose, or to prop open a fire door which should be kept closed.

■ **not to indulge in 'horseplay' or practical jokes**

The team leader sometimes has to take care that a 'harmless bit of fun' is not allowed to turn into something more dangerous. A good leader will make plain what is allowed and what isn't.

■ **to obey the rules of the organization**

People tend to break safety rules for three main reasons:

■ they aren't aware of the rules;
■ they don't see any point in the rules;
■ the rules impose conflicting restrictions, such as slowing down a process which the person wants to complete as quickly as possible: there is often a great temptation to 'cut corners'.

Activity 7

Can you think of an instance where someone in your team has been tempted to cut corners in a job, and thereby has compromised on safety? If so, describe it briefly.

_____

_____

_____

_____

_____

Depending on the kind of work you are in, you may have suggested some of the following. Sometimes there is pressure to cut corners, even from the employer – who should know better.

- Not bothering to put on protective clothing

  - 'I know I should have worn a safety helmet, but I was only going to be out in the yard for two minutes. How was I to know that it would be slippery and that I would fall and crack my head open?' (Man speaking from hospital bed.)

- Not using the right equipment

  - 'The step ladder was in use at the time, and I only wanted one item from the top shelf to finish the whole job. Now it looks like I'll be off work for three months.' (Woman on crutches.)

- Not isolating equipment before working on it

  - 'Yes, I admit that I should have checked that the electrical power was off before I asked young Peter to open the fuse-box. I was thinking about how much time the interruption was costing us. Now I'll have to live with this for the rest of my life.' (Supervisor at inquiry into fatal accident.)

- Working on, knowing the risks, and choosing to ignore them for one reason or another

  - 'The only way to get to the lift control box is to stick your head into the shaft. I suppose we should have shut down the system – but we'd been told that two people were trapped in the lift between floors. We've never had an accident till now. It was a succession of events that caused it. First of all the lift wasn't faulty at all – it was just that one of the doors wasn't shut properly. The trapped people got out, but no one told us. Then someone must have knocked down the warning notice on the ground floor, and somebody else used the lift just at the time Jim was leaning into the shaft. He didn't stand a chance when that balancing weight came down.' (Maintenance engineer talking after fatal accident.)

To sum up:

- Employees have responsibilities under HSWA:

  - to take care for their own health and safety, and that of their colleagues;
  - to co-operate in meeting the requirements of the law;
  - not to interfere with or misuse anything provided to protect their health, safety and welfare.
- People who cut corners endanger themselves and others.

Let's now return to that phrase 'as far as reasonably practicable'.

13

4 Levels of statutory duty

In law, there are three separate levels of statutory duty. From the lowest to the highest, they are:

- 'reasonably practicable' requirements;
- 'practicable' requirements;
- 'absolute' requirements.

Let's discuss what each of these means.

4.1 The duty to act in a 'reasonably practicable' manner

You will recall that a key phrase, repeated many times in the Health and Safety at Work etc. Act, is 'as far as reasonably practicable'.

To illustrate what is 'reasonably practicable' so far as health and safety is concerned, read the following case.

- Two men employed as labourers on a building site are given the job of carrying some panes of glass from one part of the site to another. After a couple of journeys, both the men complain of cuts to their hands, and request that protective gloves be provided. The supervisor tells them not to be so 'soft'. The men carry on for a while and then refuse to carry any more of the glass panes.

Activity 8

In your opinion:

	Yes	No
is the labourers' refusal to carry on working justified?	☐	☐
is the supervisor acting in a 'reasonably practicable' manner by insisting they carry on working without hand protection?	☐	☐

It seems sensible that the men should be provided with hand protection, because they were sustaining actual injuries in doing the work. The supervisor is not 'ensuring as far as reasonably practicable' the safety of the men in his employ. He should have anticipated the need for gloves and ensured they were provided and used.

This case was perhaps not too difficult to make a judgement about. Other situations may not be so straightforward. The expression 'so far as is reasonably practicable' has only acquired a clear meaning through many interpretations by the courts.

According to the Health and Safety Executive:

> **EXTENSION 3**
> This extract and the one below, is from page 30 of *Successful Health and Safety Management*, published by HSE Books

- 'To carry out a duty so far as is reasonably practicable means that **the degree of risk** in a particular activity or environment **can be balanced against the time, trouble, cost and physical difficulty of taking measures to avoid the risk**.

- If these are so disproportionate to the risk that it would be unreasonable for the persons concerned to have to incur them to prevent it, they are not obliged to do so.

- **The greater the risk the more likely it is that it is reasonable to go to very substantial expense, trouble and invention to reduce it**. But if the consequences and the extent of a risk are small, insistence on great expense would not be considered reasonable.

- It is important to remember that the judgement is an objective one and **the size or financial position of the employer are immaterial**.'

4.2 'Practicable' requirements

The phrase 'so far as is practicable' – without the qualifying word 'reasonably' – implies a stricter standard. The interpretation of the phrase given by HSE is as follows.

'This term generally embraces **whatever is technically possible** in the light of current knowledge, which the person concerned had, or ought to have had, at the time. The **cost, time, and trouble involved are not to be taken into account**.'

4.3 'Absolute' requirements

In health and safety regulations, such as those we will discuss in the next section, the words 'shall' or 'must' are used frequently. In these cases, there can be no argument or interpretation: the law **must** be obeyed.

Example

Regulation 7 (page 26) of the Health and Safety (Display Screen Equipment) Regulations 1992 states that:

1 Every employer **shall** ensure that operators and users at work in his undertaking are provided with adequate information about –

   a all aspects of health and safety relating to their workstations; and
   b such measures taken by him in compliance with his duties under regulations 2 and 3 as relate to them and their work.

In brief, for all employers:

■ 'reasonably practicable' means that the degree of risk has to be balanced against the cost, time and difficulty of taking measures to avoid the risk;
■ 'practicable' means the cost, time and difficulty are not to be considered – technical feasibility is the only consideration;
■ 'absolute' – often indicated by the word 'shall' – means that the law **must** be obeyed.

We looked at the employer's and employee's duties under HSWA, and now we turn to some important new regulations. These supplement the existing law, and in some cases replace older laws. In most cases, the regulations are absolute – they must be complied with.

5 Enforcing the law

It is the job of the Health and Safety Inspectorate to enforce the health and safety laws.

The Health and Safety Inspectorate have wide-reaching powers. They include the right to:

■ enter and inspect any premises, at any time, where it is considered that there may be dangers to health or safety;
■ be accompanied by any duly authorized person, such as a policeman or a doctor;

- enquire into the circumstances of accidents;
- require that facilities and assistance be provided by anyone able to give them;
- take statements;
- require that areas be left undisturbed;
- collect evidence, take photographs, make measurements and so on;
- take possession of articles;
- require the production of books and documents.

To enforce certain actions, an inspector can:

- issue a **prohibition notice**, which stops – with immediate effect – people from carrying on activities which are considered to involve a risk of serious personal injury;

- issue an **improvement notice**, which compels an employer to put right conditions which contravene the law, within a specified time period;

- initiate **prosecutions**, especially in the case of repeated, deliberate or severe offences.

It goes without saying that managers are expected to give their full co-operation to the enforcing authorities. The liability for personal prosecutions is very real.

An employer can appeal against an improvement notice or a prohibition notice. Here is an example of the case one company put up against a prohibition notice. The prohibition notice was issued to prevent a cutting machine being used, because a safety guard had been removed.

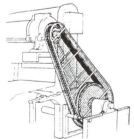

From HSE book *Work Equipment – Guidance on Regulations*, page 56.

- The guard was removed to enable the machine to cope with an oversize order which was successfully completed. When the guard was removed the electronic cut-out mechanism which would normally stop the machine running without the guard was damaged.

  The manufacturers of the electronic components for this type of guard have gone out of business, and it will take some time to find a suitable alternative, although the company is making every effort to do so.

  To have a cut-out mechanism made specially would be very expensive.

  The company appeals against the prohibition notice on the grounds of cost and difficulty.

Activity 9

3 mins

Imagine you have to make a judgement on this appeal. You understand that cost and difficulty is an important consideration for any organization, but your main concern is that of safety.

Would you agree that the prohibition should be lifted, given the circumstances?   YES/NO

Give a brief reason for your answer.

_____

_____

_____

_____

_____

_____

In spite of the cost and difficulties, there is not sufficient reason to lift the prohibition notice. Safety must come first. If the company were to use the machine without a guard, or with a guard that could be removed easily because there is no cut-out mechanism, someone might be seriously injured.

It is in fact very difficult to make a successful appeal against a prohibition note or an improvement notice.

6 Safety representatives and committees

EXTENSION 4
If you are interested in this subject, you may want to read the Health and Safety Commission booklet *Safety Representatives and Safety Committees.*

Everyone has a part to play in health and safety matters. It seems sensible for an employer, therefore, to encourage employee participation in this area.

In this section, we'll take a brief look at the role of safety representatives and safety committees in health and safety.

The regulations covering safety representatives and safety committees are included in Section 15 of HSWA.

6.1 The safety representative

A safety representative is someone appointed by a recognized trade union to represent employees on health and safety matters at work. Because he or she needs to be familiar with the hazards of the workplace and the work being done, safety representatives are usually people with two or more years' experience in that particular job.

Safety representatives have three main functions. The first one is to take all reasonably practicable steps to keep themselves informed.

Activity 10

What kind of information do you think an employees' representative on health and safety would need, in order to do a good job?

_____

_____

_____

_____

Safety representatives would surely need to be familiar with:

- what the law says about the health and safety of people at work, and particularly the people they represent;
- the particular hazards of the workplace;
- the measures needed to eliminate these hazards, or to cut down the risk from them;
- the employer's health and safety policy, and the organization and arrangements for putting that policy into practice.

The second main function is to encourage co-operation between their employer and employees so that:

- measures can be developed and promoted to ensure the health and safety of employees;
- the effectiveness of these measures can be checked.

The third function is to bring to the attention of the employer any unsafe or unhealthy conditions or working practices, or unsatisfactory welfare arrangements.

Activity 11

4 mins

Knowing the functions of a safety representative, you may be able to work out the kind of activities involved. Jot down **two** possible activities, if you can.

_____

_____

_____

_____

As you may have mentioned, safety representatives will usually be involved in:

- talking to employees about particular health and safety problems;
- carrying out inspections of the workplace to see whether there are any real or potential hazards which haven't been adequately addressed;
- reporting to employers about these problems and other matters connected to health and safety in that workplace;
- taking part in accident investigations.

Inspections and reports should be recorded formally in writing.

6.2 Safety committees

An employer is legally obliged to set up a safety committee after receiving written requests to do so from two safety representatives.

Typically, a safety committee is composed of:

- a full-time or part-time safety officer – if there is one;
- a works engineer;
- safety representatives;
- a company doctor – if there is one;
- a senior executive of the organization.

The safety committee:

- reviews the organization's health and safety rules and procedures;
- studies statistics and trends of accidents and health problems;
- considers reports and information received from health and safety inspectors;
- keeps a watch on the effectiveness of the safety content of employee training.

Let's look at the kind of accident statistics that might typically be collected in an organization.

Activity 12

Read through the accident statistics on page 22 and try to spot **three** facts that might be of interest to a safety committee.

For instance, one such fact would be that almost all accidents were minor in nature.

_____

_____

_____

_____

_____

_____

A safety committee may have noted that:

- most accidents were minor in nature;
- the inspecting and packing of glass seems to have resulted in several injuries (though only one cut);
- two of the injuries happened during loading operations;
- people seem to spend a lot of time off work for apparently minor injuries;
- three new members of staff were involved in accidents.

This short section should have given you an idea of the functions of safety representatives and safety committees.

You may also want to note the following point of law. Under the Trade Union Reform and Employment Rights Act, 1993, all employees, regardless of their length of service, have a right to complain to an industrial tribunal if they are dismissed or otherwise victimized for:

- carrying out any health and safety activities for which they have been designated by their employer;
- performing any functions as an official or employer-acknowledged health and safety representative or safety committee member;
- bringing a reasonable health and safety concern to their employer's attention in the absence of a representative or committee who could do so on their behalf;
- leaving their work area or taking other appropriate action in the face of serious and imminent danger.

Note that HSE must be notified whenever a person at work is incapacitated for normal work for more than three days as a result of an injury caused by an accident at work. This point is dealt with again on page 50 of this workbook.

Portdown Engineering (Bosham) Ltd.
ACCIDENT STATISTICS SHEET

Period: From 23.03.97 To 19.04.97

Date of accident	Date of engagement	Name, clock no. and department	Sex	Age	Occupation	Description of accident	Nature of injury	Absent (days)	Code
25.03.97	30.09.91	J.P. Peters Clock No. 85023 Blownware Factory	F	41	Inspector	While inspecting glass in the Blownware factory a flask exploded causing injury.	Cut right forearm.	6	1/14
09.04.97	10.01.97	C.K. Rush Clock No. 91025 Packing Dept.	F	22	Packer	While packing glass and using stapler machine she felt pain in her neck.	Pain in neck.	5	14
19.04.97	26.02.92	J.C. Isoz Clock No. 86012 Packer	M	32	Packer	While packing glass he developed pains in both arm and back.	Pain in arms and back.	10	14
26.03.97	13.10.93	S.J. Ruffle Clock No. 87222 Deptford Dec' Cent.	F	43	Packer	While lifting glass from crate she strained her back.	Back strain.	15	5
23.03.97	05.03.97	I.T. Hones Clock No. 91126 Packing Dept.	M	35	Packer	While packing glass, a roll of shrink wrap material, standing on end, fell over and hit foot.	Bruised right big toe.	13	5
10.04.97	27.02.90	A.L. Carvell Clock No. 84076 Deepdale Dept.	F	37	Inspector	While opening cartons to inspect contents, she cut bend of small finger causing injury.	Very small cut to small finger of left hand.	2	1/14
03.04.97	18.07.90	J.Y. Blincowe Clock No. 84321 Plant and Services	M	62	Steel erector	While walking outside of Steel erector shop, he twisted his ankle.	Injury to right outer ankle.	9	14
23.03.97	20.02.97	D.M. Hussein Clock No. 89101 Receiving Stores	M	25	Labourer	While loading laundry onto trailer, he caught his knee on edge of trailer.	Pain in right knee.	16	4
07.04.97	23.08.85	J.Austin Clock No. 79342 Transport	M	48	Driver	While trailer was being parked at Deptford parking area, he was jammed between unit and wall.	Fractures to right collar bone, right arm and foot.	27	5
13.04.97	19.09.79	G.I.K. Shemwell Clock No. 73211 Transport	M	39	Fork lift driver	While loading pallets into trailer in Pressware Factory, he felt pain in his back.	Pain in lumbar region of back.	10	5

Self-assessment 1

1 Pick the correct statements from among the following.

a Statute law is derived from court decisions. ☐

b Contract law is relatively unimportant in health and safety matters. ☐

c Following an accident, an organization may be prosecuted under either criminal law or civil law. ☐

d European law takes precedence over UK law. ☐

e The Health and Safety at Work etc. Act is a disabling act. ☐

f The Health and Safety at Work etc. Act places an obligation on employers to take care of the health and safety of customers on its premises. ☐

g Employees have duties to co-operate with employers in meeting the requirements of the law. ☐

h If an employee is given defective equipment, and gets hurt as a result, it's entirely the employer's fault. ☐

i 'As far as reasonably practicable' means that the degree of risk can be balanced against the cost of taking measures to avoid the risk. ☐

j 'So far as is practicable' means that the degree of risk can be balanced against the cost of taking measures to avoid the risk. ☐

2 The following statements about safety representatives have some words missing. Fill in the blanks from the words listed below.

Safety representatives may be involved in:

ACCIDENT   CO-OPERATION
HAZARDS   INSPECTIONS
PRACTICES   REPORTING
TALKING   UNSAFE
WELFARE   WORKPLACE

■ _____ to employees about particular health and safety problems;

■ encouraging _____ between their employer and employees;

■ carrying out _____ of the workplace to see whether there are any

real or potential _____ that haven't been adequately addressed;

■ bringing to the attention of the employer any _____ or unhealthy

conditions or working _____ , or unsatisfactory _____

arrangements;

■ _____ to employers about these problems and other matters

connected to health and safety in that _____ ;

■ taking part in _____ investigations.

3   Match the correct description from the list on the right with each term on the left.

a   Approved codes of practice (ACOPs)

i   Acts of Parliament (such as the Health and Safety at Work etc. Act 1974), together with a great many 'statutory instruments' or 'subordinate legislation'.

b   Civil law

ii   Based on case law: the decisions made in courts over the centuries. Once a judgement is made, a precedent is established.

c   Common law

iii   Anyone committing a crime has offended against the state, and is in breach of this. If an organization fails to comply with its statutory health and safety duties, its officers may be prosecuted.

d   County courts

iv   A plaintiff sues a defendant, usually for damages, that is, financial compensation. As an example, an individual may sue an employer if he or she is injured at work.

e   Criminal law

v   Deal mainly with the less serious criminal offences.

f   Crown Courts

vi   More serious cases go to one of these, and these courts also hear appeals from other courts.

g   EU Directives

vii   They deal only with civil matters. More complex civil cases, or ones involving large sums of money, go instead to one of the three divisions of the High Court of Justice.

h   Magistrates' courts

viii   Bind member countries to comply with an agreed ruling. They are normally made into national laws by each state.

i   Statute law

ix   Issued by the Health and Safety Commission (HSC) as interpretations of regulations, and are intended to help people apply the law in practice.

Answers to these questions can be found on pages 867.

7 Summary

- English law (applicable in England and Wales) is derived from three principal sources: statute law, common law, and contract law.

- Although it is important to distinguish between civil law and criminal law, both are important to organizations in terms of health and safety.

- Following an accident, actions may be brought against an organization through the criminal courts (for breach of statutory duty); through the civil courts (for breach of common law or statutory duty); or through an industrial tribunal (for injury benefits).

- Magistrates' courts largely deal with the less serious criminal offences. More serious cases go to one of the Crown Courts, sometimes having been referred there by a magistrates' court. Crown Courts also hear appeals from magistrates' courts. County courts deal only with civil matters. More complex civil cases, or ones involving large sums of money, go instead to one of the three divisions of the High Court of Justice.

- European laws take precedence over national laws in the courts, if ever there is any conflict.

- Approved codes of practice (ACOPs) are issued by the Health and Safety Commission (HSC) as interpretations of regulations, and are intended to help people apply the law in practice. Guidance notes may include advice on action to be taken by employers in order to conform with the law.

- Both employers and employees have duties under the Health and Safety at Work etc. Act, 1974 (HSWA).

- The employer has a duty 'to ensure, as far as reasonably practicable, the health, safety and welfare at work of all his employees'. The employee has a duty to take reasonable care for the health and safety of himself and of other persons who may be affected by his acts or omissions at work.

- Every employer of five or more people must prepare and keep up to date a written statement of general policy with respect to the health and safety at work of employees, and the organization and arrangements for carrying out that policy, and to bring the statement and any revision of it to the notice of all employees.

- For all employers:
  - 'reasonably practicable' means that the degree of risk has to be balanced against the cost, time and difficulty of taking measures to avoid the risk;
  - 'practicable' means the cost, time and difficulty are not to be considered – technical feasibility is the only consideration;
  - 'absolute' – often indicated by the word 'shall' – means that the law must be obeyed.

- The Health and Safety Inspectorate have wide-reaching powers, including the right to enter and inspect any premises, at any time, where it is considered that there may be dangers to health or safety.

- A safety representative is someone appointed by a recognized trade union to represent employees on health and safety matters at work.

- An employer is legally obliged to set up a safety committee after receiving written requests to do so from two safety representatives.

Session B Some important health and safety laws

1 Introduction

We have already looked at the Health and Safety at Work etc. Act, 1974; this is still the most important piece of health and safety legislation.

However, there are a number of other statutes that are relevant to most people at work, and in this session we will discuss them.

To recap: the difference between Acts and regulations is that regulations are 'statutory instruments', containing more specific requirements than the 'framework' Acts that precede them. Since 1974, all regulations concerned with health, safety, and welfare, have been passed by Parliament in furtherance of the Health and Safety at Work etc. Act, 1974.

As a result of Directives agreed by the European Union, six new regulations were introduced in January 1993. This 'six pack' comprised the following regulations:

- the Management of Health and Safety at Work Regulations, 1992;
- the Workplace (Health, Safety and Welfare) Regulations, 1992;
- the Manual Handling Operations Regulations, 1992;
- the Health and Safety (Display Screen Equipment) Regulations, 1992;
- the Personal Protective Equipment at Work Regulations, 1992;
- the Provision and Use of Work Equipment Regulations, 1992.

The first of these was amended by the Management of Health and Safety at Work (Amendment) Regulations, 1994.

Another very significant piece of legislation is the Control of Substances Hazardous to Health Regulations (COSHH 2) 1994.

We will review all of these, and in addition, look briefly at a number of other acts.

27

2 Management of Health and Safety at Work Regulations, 1992 (MHSWR)

MHSWR applies to all kinds of work, apart from sea-going ships.

These regulations help to spell out the duties and responsibilities of employers much more explicitly than HSWA.

According to HSE:

'Their main provisions are designed to encourage a more systematic and better organized approach to dealing with health and safety.'

Specifically, MHSWR requires employers to:

1 assess the risks of the job

They must assess the risk to the health and safety of their employees and anyone else who may be affected by their activity, so that the necessary preventive and protective measures can be identified. Employers with more than five employees have to record the significant findings of the assessment.

2 implement necessary measures

Any required health and safety measures that follow from the risk assessment must then be put into practice.

EXTENSION 5
The approved code of practice, *Management of Health and Safety at Work Regulations, 1992* gives guidance on risk assessment.

Comment

To carry out a **risk assessment**, you must:

■ identify the hazard;
■ measure and evaluate the risk from this hazard;
■ put measures into place that will either eliminate the hazard, or control it.

Example

In a work process involving the hand grinding of metal, one of the hazards is noise, and this fact must be recognized. During a risk assessment, the level of noise would need to be measured, and early action taken to protect workers – such as effective ear protectors.

3   provide health surveillance

Employers have to provide appropriate health surveillance for employees, where the risk assessment shows it to be necessary.

EXTENSION 3
A practical guide for managers about both HSWA and the Management of Health and Safety at Work Regulations (MHSWR) is *Successful Health and Safety Management.*

Comment

The purpose of **health surveillance** is to:

■ identify adverse affects early, well before disease becomes obvious;
■ rectify inadequacies in control, and so reduce the risks to those affected or exposed;
■ inform those at risk, as soon as possible, of any damage to their health, so that they can take action, and perhaps change their job;
■ to reinforce health education, for example by reminding workers to use the personal protective equipment provided.

Those whose health should be closely monitored include people who:

■ work in dust-laden atmospheres;
■ handle toxic or harmful substances, such as lead, chromium or pesticides;
■ work in noisy environments;
■ use equipment or materials potentially damaging to the eyes.

4   appoint competent persons

These are people who will help the employer devise and apply the measures needed to comply with health and safety laws. They can be regarded as competent if they have 'sufficient training, and experience or knowledge and other qualities, properly to undertake' the role. Internal staff or external consultants may be used.

5   provide information

Employees, together with temporary employees and others in the employer's undertaking, must be given information they can understand, about health and safety matters.

**Portfolio
of evidence
A1.2**

Activity 13

15
mins

This Activity may provide the basis of appropriate evidence for your S/NVQ portfolio. If you are intending to take this course of action, it might be better to write your answers on separate sheets of paper.

As a team leader, you are expected to give information to your team members on their responsibilities for maintaining healthy, safe, and productive work conditions. This information should comply with your organization's requirements, and with the law.

Summarize the way that you currently go about fulfilling this responsibility.

_____

_____

_____

_____

_____

_____

Now write down the plans you have for improving the health and safety information you provide to your team.

_____

_____

_____

_____

_____

_____

Describe how you will ensure that they have understood this information.

_____

_____

_____

_____

_____

_____

To continue, MHSWR also requires employers to:

6   provide training

Employees have to be given adequate health and safety training, and the employer must ensure they are capable enough at their jobs to avoid risks.

This may well be part of your job, too.

Activity 14

3 mins

The law imposes a duty on employers to provide any necessary training on safe practices. Make a note of at least **three** aspects of safe practice in which you think the law would expect training and information to be provided.

For example: 'how to work safely in a particular job'.

_____

_____

_____

_____

_____

_____

Don't forget that, to your team you are the employer.

An employee needs to know, through clear instructions and/or training, everything that concerns personal safety, including:

■   how to work safely in his or her job;
■   what to do if something goes wrong;
■   where to find safety equipment, and how to use it;
■   all relevant legal requirements;
■   what steps he or she needs to take to safeguard the safety of others;
■   any special hazards.

MHSWR also requires employers to:

7   set up emergency procedures

8   co-operate with any other employers who share a work site

9   place duties on employees to follow health and safety instructions and report danger

10 consult employees' safety representatives and provide facilities for them

Consultation must now take place on such matters as the:

- introduction of measures that may substantially affect health and safety;
- the arrangements for appointing competent persons;
- health and safety information required by law;
- health and safety aspects of new technology being introduced to the workplace.

To repeat the main points included in MHSWR, employers must:

1 assess the risks of the job;
2 implement necessary measures;
3 provide health surveillance;
4 appoint competent persons;
5 provide information;
6 provide training;
7 set up emergency procedures;
8 co-operate with any other employers who share a work site;
9 place duties on employees to follow health and safety instructions and report danger;
10 consult employees' safety representatives and provide facilities for them.

Activity 15

3 mins

Make a note of the aspects of MHSWR you think you are most likely to be involved in, in your job as first line manager.

_____

_____

_____

_____

_____

_____

Managers at all levels may be expected to participate in implementing MHSWR. In particular, you may have noted:

- implementing specific measures, following risk assessment;
- providing information and training for your team;
- helping to set up emergency procedures.

2.1 The Management of Health and Safety at Work (Amendment) Regulations, 1994

These amendments to MHSWR were introduced as a result of the European Directive on Pregnant Workers. They require employers to:

■ assess the risks to the health and safety of women who are pregnant, who have recently given birth, or who are breastfeeding;
■ ensure that workers are not exposed to risks identified by the risk assessment, which would present a danger to their health and safety.

Now for the other five regulations in the 'six-pack'. You don't have to remember everything we discuss, but you may want to:

■ take a note of those aspects that you feel are particularly relevant to your job and position;
■ decide whether to follow up the references, so that you can find out more.

3 Workplace (Health, Safety and Welfare) Regulations, 1992 (WHSWR)

These regulations replace a total of thirty-eight items of older legislation. They cover many aspects of health, safety and welfare in the workplace, and apply to most places of work.

They do **not** apply to:

■ ships and boats;
■ building operations or works of engineering construction;
■ mines and mineral exploration sites;
■ work on agricultural or forestry land away from main buildings.

WHSWR stipulates general requirements for working conditions, related to:

1 the working environment – including:

- ☐ temperature;
- ☐ ventilation;
- ☐ lighting;
- ☐ room dimensions;
- ☐ workstations and seating;
- ☐ outdoor workstations, such as weather protection.

Are you obeying the law?

As you read through these headings, tick the boxes of any subjects you feel are particularly relevant to your own circumstances, and that you would like to find out more about.

Example

The regulations say that:

'Effective and suitable provision shall be made to ensure that every enclosed workspace is ventilated by a sufficient quantity of fresh or purified air.'

In offices and many other workplaces, windows will provide enough ventilation. Alternatively, air conditioning systems may be installed.

Comment

Many of the requirements in these regulations appear to be 'common sense', and it is probably true to say that most workplaces are properly ventilated, heated and so on. Nevertheless, employers must obey the law.

---

2 safety – including:

EXTENSION 6
If you want to learn more, the approved code of practice: *Workplace Health, Safety and Welfare* is available from HSE Books. You'll find it listed on page 83.

- ☐ safe passage of pedestrians and vehicles;
- ☐ windows and skylights (safe opening, closing and cleaning);
- ☐ organization and control of traffic routes;
- ☐ glazed doors and partitions (use of safe material and marking);
- ☐ doors, gates and escalators (safety devices);
- ☐ floors (their construction and condition);
- ☐ obstructions and slipping and tripping hazards;
- ☐ falls from heights and into dangerous substances, and falling objects.

Example

The regulations say that:

1 'Every workplace shall be organized in such a way that pedestrians and vehicles can circulate in a safe manner.
2 Traffic routes in a workplace shall be suitable for the persons or vehicles using them, sufficient in number, in suitable positions and of sufficient size.'

The code of practice reminds us that: 'In some situations, people in wheelchairs may be at greater risk than people on foot, and special consideration should be given to their safety.'

---

3   **welfare facilities** – including:

- ☐ toilets;
- ☐ washing, eating and changing facilities;
- ☐ provision of drinking water;
- ☐ clothing storage, and facilities for changing clothing;
- ☐ seating;
- ☐ rest areas (and arrangements in them for non-smokers);
- ☐ rest facilities for pregnant women and nursing mothers.

Example

The regulations say that:

'An adequate supply of drinking water shall be provided for all persons at work in the workplace.'

---

4   **housekeeping** – including:

- ☐ maintenance of workplace, equipment and facilities;
- ☐ cleanliness;
- ☐ removal of waste materials.

Activity 16

3 mins

Make a note here if you want to find out more, and of the action you intend to take. (One way is to get hold of a copy of the Approved Code of Practice, listed in Extension 5, but you may be able to get the information through your organization. If you can get help with the interpretation of the regulations, you should certainly do so.)

_____

_____

_____

_____

_____

_____

4 Manual Handling Operations Regulations, 1992 (MHOR)

EXTENSION 7
The Manual Handling Operations Regulations, 1992 are explained in the HSE booklet *Manual Handling – Guidance on Regulations.*

These regulations cover the lifting and manoeuvring of loads of all types. They require the employer to:

■ consider whether a load must be moved, and if so, whether it could be moved by non-manual methods;

■ assess the risk in manual operations and (unless it is very simple) make a written record of this assessment;

■ reduce the risk of injury as far as is reasonably practicable.

The following is a summary of the questions that should be asked regarding four aspects of a manual handling operation.[2]

Making an assessment: some important questions

1 The task

■ Is the load held or manipulated at a distance from the trunk, so increasing the stress on the lower back?

■ Does the task involve:
   – twisting the trunk?
   – stooping?
   – reaching upwards?
   – excessive lifting or lowering distances?
   – excessive carrying distances?
   – pushing or pulling of the load?
   – a risk of sudden movement of the load?
   – frequent or prolonged physical effort?
   – insufficient rest or recovery periods?
   – a rate of work imposed by a process?

2 The load

■ Is the load:
   – heavy?
   – bulky or unwieldy?
   – difficult to grasp?
   – unstable, or are its contents likely to shift?
   – sharp, hot or otherwise potentially damaging?

3 The working environment

■ Are there:
   – space constraints preventing good posture?
   – uneven, slippery or unstable floors?
   – variations in the level of floors or work surfaces?
   – extremes of temperature or humidity?
   – ventilation problems or gusts of wind?
   – poor lighting conditions?

4 Individual capability

■ Does the task require unusual strength, height etc.?
■ Does the job put at risk those who might be pregnant or have a health problem?
■ Does the task require special information or training for its safe performance?

---

[2] Adapted from *Manual Handling – Guidance on Regulations* published by HSE, pages 12–20.

Activity 17

3 mins

	Yes	No
Does your team's job involve manual handling?	☐	☐
Are you confident that you are managing manual handling tasks efficiently and effectively?	☐	☐

If not, how will you learn more?

_____

_____

_____

5 Health and Safety (Display Screen Equipment) Regulations, 1992

EXTENSION 8
These regulations are explained in *Display Screen Equipment Work — Guidance on Regulations* published by HSE.

These regulations put into law the employer's duties regarding the operation of display screen equipment by employees. Employers have to:

- assess and reduce the risks from display screen equipment;
- make sure that workstations satisfy minimum requirements;
- plan to allow breaks or change of activity;
- provide information and training for users;
- give users eye and eyesight tests and (if need be) special glasses.

Display screen equipment includes cathode ray tubes (CRTs), liquid crystal displays, and any other technology.

The regulations do not apply to:

- drivers' cabs or control cabs for vehicles and machinery;
- display screen equipment on board a means of transport;
- display screen equipment mainly intended for public operation;
- portable systems not in prolonged use;
- calculators, cash registers or any equipment having a small data or measurement display required for direct use of the equipment;
- window typewriters.

The figure, and the text below it, summarize the minimum requirements for workstations.[3]

Key to illustration:

1   Adequate lighting

2   Adequate contrast, no glare or distracting reflections

3   Distracting noise minimized

4   Leg room and clearances to allow postural changes

5   Window covering

6   Software: appropriate to task, adapted to user, provides feedback on system status, no undisclosed monitoring

7   Screen: stable image, adjustable, readable, glare/reflection free

8   Keyboard: usable, adjustable, detachable, legible

9   Work surface: allow flexible arrangements, spacious, glare free

10   Work chair: adjustable

11   Footrest

[3] Figure and text from *Display Screen Equipment Work – Guidance on Regulations* published by HSE.

Activity 18

3 mins

	Yes	No
Does your team's job involve working with display screen equipment for long periods?	☐	☐
Are you confident that you comply with the law?	☐	☐

If not, how will you learn more?

_____

_____

_____

_____

_____

_____

6 Personal Protective Equipment at Work Regulations, 1992 (PPEWR)

EXTENSION 9
Further information about the PPE Regulations is given in the HSE booklet *Personal Protective Equipment at Work — Guidance on Regulations.*

Personal protective equipment (PPE) includes eye, foot and head protection equipment, safety harnesses, life jackets and so on. Employers have to:

- ensure this equipment is suitable and appropriate;
- maintain, clean and replace it;
- provide storage for it when not in use;
- ensure that it is properly used;
- give employees training, information and instruction in its use.

Complete the following Activity, to help you decide what further action you need to take regarding PPE.

Activity 19

10 mins

If your team needs to use personal protective equipment, or you think they may need to use it, answer the questions below.

- Is the equipment appropriate to the risks, and to the conditions at the place where exposure might occur? ☐

- Does it take account of ergonomic requirements, and the state of health of the persons who may wear it? ☐

- Does it fit the wearer? ☐

- Is it effective in preventing or controlling the risk? ☐

- Is it compatible with other equipment? ☐

- Have all the following risks been assessed?

  Head injury? ☐   Eye injury? ☐

  Face injury? ☐   Inhalation of airborne contaminants? ☐

  Noise-induced hearing loss? ☐   Skin contact? ☐

  Bodily injury? ☐   Hand or arm injury? ☐

  Leg or foot injury? ☐   Vibration-induced injury? ☐

- Is appropriate accommodation provided for the PPE when it isn't in use? ☐

- Have the users been given adequate information, instruction and training? ☐

- Do all the users:

  - use the equipment in accordance with their training and instructions? ☐

  - return the PPE to its accommodation after use? ☐

  - understand the need to report losses or defects? ☐

What further actions do you intend to take about PPE?

_____

_____

_____

_____

_____

_____

_____

_____

7 Provision and Use of Work Equipment Regulations, 1992 (PUWER)

EXTENSION 10
The Provision and Use of Work Equipment Regulations, 1992 are explained in the HSE booklet *Work Equipment – Guidance on Regulations.*

The definition of 'work equipment' is very wide, and includes a butcher's knife, a combine harvester and even a complete power station. Employers must:

- take into account working conditions and hazards when selecting equipment;
- ensure equipment is suitable for use, and is properly maintained;
- provide adequate instruction, information and training.

The next Activity is a very brief summary checklist. Complete it to help you decide what action you might want to take in regard to work equipment.

Activity 20

6 mins

Identify an important item of equipment that your team works with. Then answer the following questions about it.

Item: _____

- Is the equipment suitable for the use it is put to? ☐
- Is it well and regularly maintained? ☐
- Are adequate information and instructions available to potential users? ☐
- Have the users received adequate training in its use? ☐
- Have you provided protection against dangerous parts of the equipment, such as guards around rotating parts? ☐
- Have measures been taken to eliminate or control risks associated with the use of the equipment? ☐
- Are stop controls provided, that are easy to reach, and well marked? ☐
- Are control systems (if any) safe? ☐
- Is isolation provided from sources of energy? ☐
- Is the equipment stable? ☐
- Is it well lit, and marked with appropriate signs and warnings? ☐

Do you need to find out more about the use of work equipment, or these regulations? Write down any actions you intend to take.

_____

_____

_____

_____

_____

_____

So much for the six regulations that form the 'six pack'. Now we will look at another very important set of regulations.

8 The COSHH 2 regulations

COSHH 2 doesn't cover asbestos, lead, materials producing ionizing radiation and substances below ground in mines, which all have their own legislation.

When they were first introduced in 1988, the Control of Substances Hazardous to Health Regulations (COSHH) introduced a new legal framework for controlling hazardous substances at work. The 1994 regulations updated and replaced the 1988 version, and are known as COSHH 2.

The COSHH 2 regulations cover virtually all substances that can affect health.

Activity 21

3 mins

Can you think of any substances which could possibly affect the health of people where you work?

_____

_____

_____

_____

There are over 40,000 substances which are classed as hazardous. Wherever you work – in a factory, in a workshop, in a quarry, on a farm or garden, in a laboratory, a storage area, or even in an office – **hazardous substances probably exist**. They may include:

- anything brought into a workplace to be worked on, used or stored: these may include corrosives, acids or solvents used in cleaning materials;
- dust and fumes given off by a work process;
- finished products or residues from a work process.

Anything very toxic, toxic, corrosive, harmful or irritant comes under COSHH 2. Examples are chemicals, agricultural pesticides, wood treatment chemicals, dusts and substances containing harmful micro-organisms. By law, the containers of hazardous substances must be labelled as being hazardous, and they must state what the hazard is.

Example of label

Toxic or very toxic

8.1 Employers' duties under COSHH 2

Under the COSHH 2 regulations, employers have to:

- **determine the hazard** of substances used by the organization;
- **assess the risk** to people's health from the way the substances are used;
- **prevent anyone being exposed** to the substances, if possible;
- if exposure cannot be prevented, decide how to **control the exposure** so as to **reduce the risk**, and then establish effective controls;
- ensure that the controls are **properly used and maintained**;
- **examine and test the control measures**, if this is required;
- **inform, instruct and train** employees (and non-employees on the premises), so that they are aware of the hazards and how to work safely;
- if necessary, **monitor the exposure** of employees (and non-employees on the premises), and provide **health surveillance** to employees if necessary.

Example of label

Corrosive

8.2 Your job and COSHH 2

Now let us consider what impact COSHH 2 has on your job.

Activity 22

5 mins

From what you've read so far, can you suggest how first line managers and team leaders can play a part in helping their employer comply with the COSHH regulations, and so reduce the risk to employees from hazardous substances? Try to list **two** or **three** positive actions that might be taken by someone in your position.

_____

_____

_____

Example of label

Harmful or irritant

- Probably the most important role first line managers and team leaders can play is in informing and training the workteam about the hazards and the correct procedures to be followed. We'll look at the kind of information and training needed, in a moment.
- They can and should ensure that safety procedures are followed, and set a personal example in following the correct procedures, consistently and carefully.
- Team leaders are also usually in a good position to assess the likely behaviour of people when they have to deal with hazardous situations. If so, they can assist their employer in determining what people do and what they might do.
- If protective clothing and/or emergency facilities are provided, it is often the team leader's job to ensure that these items are available when they're needed, and are properly maintained.
- Leaders should ensure that only those substances are used whose risks have been assessed, and that team members have been trained in the safe handling of these substances.

Let's now look at the aspects of training and providing information to staff about hazardous substances.

A team leader will need to make sure of the following.

- Team members should understand the hazards.

    All suppliers are compelled by law to label hazardous substances. Everyone using such a substance needs to be trained to read and understand container labels and to follow the supplier's advice. More detailed advice is usually available on suppliers' 'data sheets', which list all the hazards of a substance, the effects of exposure and methods of treatment.

- Team members should understand how risks are controlled.

    Procedures for controlling risks must be clearly laid down.

- Team members should understand the precautions they have to take.

    Precautions and procedures will almost certainly need to be demonstrated.

- Team members should understand what to do in case of emergency.

    Emergency procedures need to be demonstrated and practised.

Activity 23

6 mins

This Activity may provide the basis of appropriate evidence for your S/NVQ portfolio. If you are intending to take this course of action, it might be better to write your answers on separate sheets of paper.

There are few workplaces where hazardous substances do not exist. If you have hazardous substances in your work area, what training does your team receive about them? Summarize the kind of training they get.

_____

_____

_____

_____

_____

_____

Based on the points listed above, write down your plans for improving their training in this important area.

_____

_____

_____

_____

_____

_____

What further actions do you think you need to take, in order to comply with COSHH 2? (You may decide to find out more about the regulations before you answer this.)

_____

_____

_____

_____

_____

_____

EXTENSION 11
In the Extension you will find reference to the HSE ACOP on COSHH 2.

Now we have covered the 'six-pack' regulations, and looked at COSHH 2, we move on to some other important health and safety laws.

9 Other laws

EXTENSION 12
The address and telephone number of the Health and Safety Executive is given in this extension.

Although we haven't room to cover them in detail in this unit, there are other laws which you should be aware of. Those that we will look at briefly are:

- the Factories Act, 1961;
- the Offices, Shops and Railway Premises Act, 1963;
- the Electricity at Work Regulations, 1989;
- the Health and Safety (First Aid) Regulations, 1981;
- the Noise at Work Regulations, 1989;
- the Reporting of Injuries, Diseases and Dangerous Occurrences Regulations, 1995.

It would not be surprising if you are feeling rather overwhelmed by this long list of legislation we are working through. To repeat the point made at the start of the workbook: don't feel you have to remember everything. Just continue to read it through, and complete the activities. Make a note to find out more about those laws that are especially relevant to you.

9.1 The Factories Act, 1961

At the time of writing, a new set of regulations affecting young people had been proposed, but had not yet been made law: the Health and Safety (Young Persons) Regulations.

Much of the Factories Act, 1961 has been replaced by the legislation we have already looked at: the Workplace (Health, Safety and Welfare) Regulations, 1992 (WHSWR), and the Provision and Use of Work Equipment Regulations, 1992 (PUWER). The provisions remaining are listed below. Again, tick any section you'd like to find out more about.

EXTENSION 1
The book *Health and Safety Law*, by Jeremy Stranks, is a good source of reference for this act, and for most other health and safety legislation.

☐ Cleaning of machinery by young people (Section 20)

This places restrictions on the cleaning work that young people (of 16–18 years of age) can be given to do.

☐ Training and supervision of young people working at dangerous machines (Section 21)

Young people must be trained and supervised if they are to work at dangerous machines.

☐ Hoists and lifts (Sections 22 and 23)

These sections of the Factories Act, set out general provisions related to hoists and lifts.

☐ Chains, ropes and lifting tackle (Section 26)

This section relates to the soundness, load rating, certification, annealing, and registering of chains, ropes, and lifting tackle.

☐ Cranes and other lifting machines (Section 27)

This section requires that all parts and working gear of this type of machinery shall be of good construction, of sound material, of adequate strength, and be free from defect.

☐ Dangerous fumes and lack of oxygen (Section 30)

This relates to hazards existing in confined places, and the precautions that should be taken.

☐ Precautions with respect to explosive or inflammable dust, gas, vapour or substance (Section 31)

Explosions may occur during certain work processes, including grinding and sieving, due to an accumulation of dust, and the section deals with this hazard, and protection against it. It also covers explosive or inflammable gases and vapours used in some plants.

☐ Steam boilers (restrictions on entry) (Section 34)

This section forbids entry to steam boilers unless certain safety precautions are taken.

☐ Removal of dust or fumes (Section 63)

This requires that all practicable measures be taken to protect people against the inhalation of dust or fumes likely to be injurious.

9.2 The Offices, Shops and Railway Premises Act, 1963

Only three sections of this act remain in force. They are:

☐ Exposure of young persons to danger in cleaning machinery (Section 18)

No young person must clean any machinery, if they put themselves at risk by doing so.

☐ Training and supervision for working at dangerous machines (Section 19)

No person must work at a dangerous machine without adequate training and supervision.

☐ Notification (Section 49)

Before work commences in offices or shops, the employer must notify the enforcing authority (the local authority) on Form OSR1.

9.3 The Electricity at Work Regulations, 1989

The purpose of these regulations is to require that precautions be taken to prevent the risk of death or injury from electricity at work. They set out general principles for electrical safety, rather than specifying detailed requirements.

Subjects covered include:

- earthing or other suitable precautions;
- the means of protecting from excess current;
- the means for cutting off the supply and for isolation;
- people's competence to prevent danger or injury.

9.4 The Health and Safety (First Aid) Regulations, 1981

Under this regulation, all employers in the UK are required to:

- provide first aid

Equipment and facilities must be provided that are 'adequate and appropriate' in the circumstances for enabling first aid to be rendered to employees if they are injured or become ill at work.

- inform employees of first aid arrangements.

9.5 The Noise at Work Regulations, 1989

Noise causes hearing damage, and the damage is accumulative: the longer you are exposed to excessive levels of noise, the more likely you are to suffer hearing loss.

These regulations 'are intended to reduce hearing damage caused by loud noise'. They specify three 'action levels:

- at the 'First Action Level' (measured as 85 dB(A)), employers must:
  - get noise assessed by a competent person, and keep a record;
  - inform employees about the risks to their hearing;
  - provide ear protectors if requested;
  - advise any employee who thinks their hearing is being affected to seek medical advice.

As a rough check, at this noise level it begins to be difficult to hear what someone is saying when they are two metres away.

■ at the second and peak action levels (measured as 90 db(A) and 140 dB respectively), the noise exposure must be controlled, preferably by reducing the noise, and as a last resort by providing ear protectors.

The diagram shows typical noise levels associated with work activities. The bands show the length of time that workers can be exposed to such noise before their 'noise dose' exceeds the Action Levels.

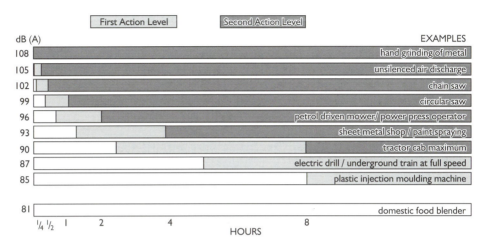

Redrawn from *Essentials of Health and Safety at Work*, published by HSE. Reproduced with kind permission of HMSO.[4]

Activity 24

According to this diagram, approximately how much time should be allowed to lapse before someone exposed to the noise of a chain saw:

a is informed about the risks to their hearing; _____

b is provided with ear protectors: _____

The best answer, in both cases, is 'no time at all'. According to the diagram, a chain saw is likely to be around 102 dB(A). This is above the First Action Level of 85 db(A), so a known risk exists, and after only a very short time the 'noise dose' will have exceeded that advisable at this level. After around half an hour, the second action level is invoked. Having this information, anyone working with a chain saw should know to wear ear protectors all the time.

---

[4] Crown copyright 1990

9.6 The Reporting of Injuries, Diseases and Dangerous Occurrences Regulations, 1995

Whenever any of the following events occurs, it must be reported in writing to the enforcing authority (usually the HSE). In addition, an event of type 1, 2 or 3 must first of all be notified to the enforcing authority 'by quickest practicable means'.[5]

1 the **death** of any person as a result of an accident arising out of or in connection with work;

2 any person at work suffering any of **certain injuries** or conditions as a result of an accident arising out of or in connection with work, including:

- **fracture of the skull, spine or pelvis**;
- **fracture of any bone** except in the hand or foot;
- **amputation** of: a hand or foot; or a finger, thumb or toe;
- **loss of the sight** of an eye or serious injury to the eye;
- injury as a result of **electric shock**;
- **loss of consciousness resulting from lack of oxygen**;
- acute illness requiring medical treatment, or loss of consciousness, resulting from **absorption of any substance by inhalation, ingestion or through the skin**;
- acute illness requiring medical treatment resulting from **exposure to a pathogen** (such as a bacterium or virus) **or infected material**;
- any other **injury that results in the person injured being admitted immediately into hospital for more than twenty-four hours**.

3 any **dangerous occurrence**, such as an overturned crane or burst pressure vessel;

4 an employee or other person at work **being incapacitated for normal work for more than three days** as a result of an injury caused by an accident at work;

5 the **death of an employee** if this occurs some time after the reportable injury that led to the employee's death, but not more than one year afterwards.

The relevant HSE forms are F2508 (Report of an injury or dangerous occurrence) and F2508A (Report of a case of a disease).

> **EXTENSION 16**
> *A Guide to RIDDOR 95* is available from HSE Books. It gives a full list of 'dangerous occurrences'.

> **EXTENSION 17**
> Electronic versions of these forms are available from HSE.

---

[5] This section (from 'Whenever any of the following events occurs' to 'but not more than one year afterwards') adapted from HSE Booklet HSE 11 (Rev) 5/86 © Crown copyright

9.7 Industry-specific legislation and help

EXTENSION 12
The addresses and telephone numbers for HSE Books and HSE InfoLine are given on page 84.

The health and safety laws we have looked at are applicable to most places of work. Other regulations are aimed specifically at certain industry sectors, and you will need to follow up any that are relevant to your area of work. Some examples are:

- Dangerous Substances in Harbour Area Regulations, 1987;
- Docks Regulations, 1988;
- Loading and Unloading of Fishing Vessels Regulations, 1988;
- Mines (Safety of Exit) Regulations, 1988;
- Notification of New Substances Regulations, 1982;
- Offshore Installations and Pipeline Works (First Aid) Regulations, 1989;
- Quarries (Explosives) Regulations, 1988;
- Road Traffic (Carriage of Dangerous Substances in Road Tankers and Tank Containers) Regulations, 1992.

HSE also gives specific guidance on the way that general legislation applies to specific industries.

Self-assessment 2

20 mins

1 Match each requirement under the Management of Health and Safety at Work Regulations, 1992 (MHSWR) on the left with the correct comment on the right.

Under MHSWR, employers must:

a provide risk assessment;

b provide health surveillance;

c appoint competent persons;

d consult employees' safety representatives.

This includes the process of:

i identifying adverse affects; rectifying inadequacies in control; informing those at risk of any damage to their health; reinforcing health education.

ii identifying measures that may substantially affect health and safety; identifying health and safety aspects of new technology; discussing these with the relevant people.

iii identifying the hazard; measuring and evaluating the risk from this hazard; putting measures into place that will either eliminate the hazard, or control it.

iv identifying those with sufficient training, and experience or knowledge and other qualities; requiring them to devise and apply the measures needed to comply with health and safety laws.

2 For each activity on the left, identify **one** regulation that will apply to it, taken from the list on the right.

Activity

i Loading bags of flour.

_____

ii Supervising telesales.

_____

iii Running an electrical department in a superstore.

_____

iv Training fork-lift truck drivers in a fuel depot.

_____

v Supervising an area where there are high levels of dust.

_____

vi Supervising on a building site.

_____

Regulation

a Management of Health and Safety at Work Regulations, 1992 (MHSWR)

b Workplace (Health, Safety and Welfare) Regulations, 1992 (WHSWR)

c Manual Handling Operations Regulations, 1992 (MHOR)

d Health and Safety (Display Screen Equipment) Regulations, 1992

e Personal Protective Equipment at Work (PPE) Regulations, 1992 (PPEWR)

3 Fill in the blanks in the following statements with suitable words, taken from the list below.

CONTROL   INSTRUCT   SAFELY
CONTROLS   MAINTAINED   SUBSTANCES
EXPOSED   MONITOR   SURVEILLANCE
HAZARD   RISK   TEST
HEALTH

Under the COSHH 2 regulations, employers have to:

■ determine the _____ of _____ used by the organization;

■ assess the _____ to people's health from the way the substances are used;

■ prevent anyone being _____ to the substances, if possible;

■ if exposure cannot be prevented, decide how to _____ the exposure so as to reduce the risk, and then establish effective _____ ;

■ ensure that the controls are properly used and _____ ;

■ examine and _____ the control measures, if this is required;

- inform, _____ and train employees (and non-employees on the premises), so that they are aware of the hazards and how to work _____ ;

- if necessary, _____ the exposure of employees (and non-employees on the premises), and provide _____ _____ to employees if necessary.

Answers to these questions can be found on pages 87–8.

10 Summary

- The Management of Health and Safety at Work Regulations, 1992 (MHSWR) are designed to encourage a more systematic and better organized approach to dealing with health and safety.

- Under MHSWR, employers must:

  - assess the risks of the job;
  - implement necessary measures;
  - provide health surveillance;
  - appoint competent persons;
  - provide information and training;
  - set up emergency procedures;
  - co-operate with any other employers who share a work site;
  - place duties on employees to follow health and safety instructions and report danger;
  - consult employees' safety representatives and provide facilities for them.

- The Management of Health and Safety at Work (Amendment) Regulations, 1994 require employers to assess the risks to the health and safety of women who are pregnant, have recently given birth, or who are breastfeeding; and ensure that workers are not exposed to risks identified by the risk assessment, which would present a danger to their health and safety.

- The Workplace (Health, Safety and Welfare) Regulations, 1992 (WHSWR) stipulates general requirements for working conditions, related to:

  - the working environment;
  - safety;
  - welfare facilities;
  - housekeeping.

- The Manual Handling Operations Regulations, 1992 (MHOR) require the employer to:

  - consider whether a load must be moved, and if so, whether it could be moved by non-manual methods;
  - assess the risk in manual operations and (unless it is very simple) make a written record of this assessment;
  - reduce the risk of injury as far as is reasonably practicable.

- The Health and Safety (Display Screen Equipment) Regulations, 1992 require employers to:

  - assess and reduce the risks from display screen equipment;
  - make sure that workstations satisfy minimum requirements;
  - plan to allow breaks or change of activity;
  - provide information and training for users;
  - give users eye and eyesight tests and (if need be) special glasses.

- The Personal Protective Equipment at Work (PPE) Regulations, 1992 (PPEWR) require employers to:

  - ensure PPE equipment is suitable and appropriate;
  - maintain, clean and replace it;
  - provide storage for it when not in use;
  - ensure that it is properly used;
  - give employees training, information and instruction in its use.

- The Provision and Use of Work Equipment Regulations, 1992 (PUWER) require employers to:

  - take into account working conditions and hazards when selecting equipment;
  - ensure equipment is suitable for use, and is properly maintained;
  - provide adequate instruction, information and training.

- Under the Control of Substances Hazardous to Health Regulations, 1994 (COSHH 2) regulations, employers have to:

  - determine the hazard of substances used by the organization;
  - assess the risk to people's health from the way the substances are used;
  - prevent anyone being exposed to the substances, if possible;
  - if exposure cannot be prevented, decide how to control the exposure so as to reduce the risk, and then establish effective controls;
  - ensure that the controls are properly used and maintained;
  - examine and test the control measures, if this is required;
  - inform, instruct and train employees (and non-employees on the premises), so that they are aware of the hazards and how to work safely;
  - if necessary, monitor the exposure of employees (and non-employees on the premises), and provide health surveillance to employees if necessary.

■ Other laws we looked at were:

- the Factories Act, 1961;
- the Offices, Shops and Railway Premises Act, 1963;
- the Electricity at Work Regulations, 1989;
- the Health and Safety (First Aid) Regulations, 1981;
- the Noise at Work Regulations, 1989;
- the Reporting of Injuries, Diseases and Dangerous Occurrences Regulations, 1995.

Session C The law on the environment

1 Introduction

What do we mean by the environment? Albert Einstein is said to have defined it as 'everything that isn't me'. The law is a little more specific: the Environmental Protection Act, 1990 expressed it as 'all, or any, of the following media, namely air, water and land'.

Fairly obviously, health and safety and the environment are closely linked. The air we breathe, and the water we drink, must be pure if we are to remain healthy. If the environment becomes polluted, the health of human beings is also likely to suffer.

This session consists of a brief review of environmental law.

2 Background to environmental law

Though British legislation on pollution goes back to the Middle Ages, the first significant law dates from the Industrial Revolution: the Public Health Act, 1875. Legislation passed since then still has an impact on environmental law today.

2.1 A brief history of environmental law

Another notable landmark was the formation of the Alkali Inspectorate, which was formed in 1863 in order to control emissions into the atmosphere caused by the caustic soda industry. It was in fact the world's first national pollution control agency. However, in the past most public health and environmental protection was carried out locally rather than nationally.

Also, until fairly recently, laws tended to be passed in response to particular problems, rather than being planned so as to take care of the environment as a whole: they were reactive, rather than proactive. An example of this was the Deposit of Poisonous Wastes Act, 1972, which was a reaction to the much-reported fly-tipping of poisonous waste near a school playground, and which was approved by Parliament in only a few days.

57

The truth is that the environment has only been politically important during the past twenty years or so. Prior to that, governments tended not to devote much parliamentary time to the subject, because there was comparatively little pressure to do so.

As Simon Ball and Stuart Bell say in their book *Environmental Law*:

'One effect of this long, and unplanned, history is that modern Britain has inherited a far less coherent system of pollution control than many other countries.'[6]

2.2 Sources of environmental law

Unlike health and safety, very little case law has any relevance to environmental law, which is virtually all statutory.

However, like health and safety, environmental law typically consists of framework Acts, which give rise to regulations that spell out the detail. An example is the Environmental Protection Act, 1990, which made it compulsory for authorization to be obtained before 'prescribed processes' could be carried on by industry. The actual processes were not mentioned in the Act, but were listed in the Environmental Protection (Prescribed Processes and Substances) Regulations, 1991.

Activity 25

2 mins

Statute law is passed by the members of the Houses of Parliament. What body outside this country would you expect to have a considerable influence on environmental law?

_____

_____

Thinking back to the sources of health and safety law, you might have guessed that the European Union would also be of enormous importance in the environmental field.

As in other fields, EU environmental law is normally introduced to member countries by means of directives, which require national laws to be passed to bring them into line with EU law. For example, the Wildlife and Countryside Act, 1981 was introduced so as to comply with an EU Directive on wild birds; at the same time, the opportunity was taken to change a number of other areas of the law.

[6] Blackstone Press Ltd, 1991, page 8. 9–15 Aldine Street, London W12 8AW.

Some further examples of subjects on which EU Directives have been issued include:

- quality standards for water;
- quality standards for air;
- noise standards;
- lead in petrol;
- the storage and use of hazardous materials.

The environmental policies of the EU have already greatly influenced British environmental legislation, and are likely to continue to do so in the future. The basic principles of these policies are that:

- **preventative action is to be preferred** to remedial measures;
- environmental **damage should be rectified at source**;
- **the polluter should pay** for the costs of the measures taken to protect the environment;
- environmental policies should form a component of the EU's other policies.

The first point of policy – that preventative action is to be preferred to remedial measures – contrasts with the traditional approach, in which governments tended to react to problems rather than trying to prevent them.

The 'polluter pays' principle is also a change to previous practice. What it means is that producers of goods or services should be responsible for the costs of preventing or dealing with any pollution that they may cause. This includes both direct costs in the form of recompense to people and repairs to property, and environmental costs.

This does **not** mean is that it is acceptable for organizations to pollute the environment, provided they pay for clearing up the mess afterwards!

What about outside Europe?

Activity 26

Many people believe that other states should not be allowed to interfere in the affairs of the UK. Citizens of other countries sometimes have the same kind of nationalist outlook. But do you think it is possible to take this attitude when it comes to the environment? Should we look after our own problems of pollution, and ignore everyone else?

_____

_____

_____

Fairly obviously, the environment has no national boundaries, and pollution cannot always be contained within one country. We were reminded of this fact in a rather grim fashion, following the Chernobyl disaster in 1986.

■ Chernobyl is 130 km north of Kiev, in central Ukraine. On 26 April, 1986, a nuclear power plant there went out of control and caused the worst reactor disaster the world has known to date. During a period of maintenance, an experiment was conducted with the water-cooling system turned off, which led to an uncontrolled reaction. The subsequent explosion of steam blew off the reactor's protective covering, releasing around 100 million curies of radio-nuclides into the atmosphere. The radiation spread across northern Europe and some of it settled in the UK, contaminating large areas of land.

All countries need to co-operate on environmental matters, and international law, which governs relations between countries, also influences UK environmental law. For example, the North Sea Conferences had a substantial impact on the law related to the dumping of sewage sludge in the North Sea.

2.3 Areas of environmental law

There are four principal areas of environmental law:

■ pollution;
■ water;
■ nature conservation;
■ town and country planning.

3 Recent legislation

The Environmental Protection Act, 1990 (EPA) brought in some fundamental changes with regard to the control of pollution and the protection of the environment.

3.1 The Environmental Protection Act, 1990

EXTENSION 1
Further details of the Environmental Protection Act, 1990, and the Environment Act, 1995, are discussed in the book *Health and Safety Law.*

Specifically, this Act covers:

■ **air pollution** (but not vehicle emissions);
■ **waste management and disposal**;
■ **integrated pollution control (IPC)**; (we will discuss integrated pollution control shortly);
■ litter;
■ the environmental impact of genetically modified organisms;
■ noise;
■ statutory control of environmental nuisances.

3.2 The Environment Act, 1995

The Environment Act, 1995 created the **Environment Agency** for England and Wales. This body came into being on 1 April, 1996, taking over the powers of HM Inspectorate of Pollution, the National Rivers Authority, and local waste regulation authorities. In Scotland, the equivalent body is the Scottish Environment Protection Agency.

The Environment Act also:

■ established a national **strategy and framework for air quality** standards, and targets for nine types of pollutant;
■ gave new powers to local authorities to **review air quality** in their areas;
■ reinforced the '**polluter pays**' policy in respect of **contaminated land**, but recognized that land-owners should also take responsibility for some aspects;
■ made **sustainable development** a cornerstone of national waste strategies, which means making the best possible use of unavoidable waste, and minimizing the risk of pollution or harm to health arising from waste disposal or recovery;
■ introduced regulations to impose **producer responsibility** to increase the re-use, recovery or recycling of any product or material;
■ made water companies responsible for the **efficient use of water** by their customers;
■ gave the Environmental Agency powers to require action to **prevent water pollution**, and to require polluters to clean up after pollution incidents.

3.3 Other recent legislation

Most of the legislation on water pollution is now contained in the Water Resources Act, 1991.

The Wildlife and Countryside Act, 1981 includes much of the law on nature conservation.

The Town and Country Planning Act, 1990 contains most of the law on town and country planning and tree protection.

We will look into some of these acts in more detail later in Session C.

In addition to these main acts of environmental law, there is separate legislation to cover other kinds of pollution, including vehicle emissions and pesticides.

4 Integrated pollution control

The concept of **integrated pollution control (IPC)** was introduced by the Environmental Protection Act, 1990.

4.1 The objectives of IPC

The main objectives of IPC are:

- to prevent or minimize the release of prescribed substances and to render harmless any such substances that are released;
- to develop an approach to pollution control that considers discharges from industrial processes to all media in the context of the environment as a whole.

4.2 Industrial processes affected by IPC

IPC applies to any industrial process carried out in England or Wales which has been prescribed by the Secretary of State for the Environment.

The processes prescribed are listed in the Environmental Protection (Prescribed Processes and Substances) Regulations, 1991. Anyone intending to carry out one of these processes must make an application to the Environment Agency, giving full details of the process. The Environment Agency will then assess the effects on the environment as a whole, and if necessary, will require the applicant to modify the process, making it less polluting.

Before authorization is granted, various statutory objectives have to be met. Account has to be taken of emissions to all three media: land, water and air.

The Environmental Protection (Applications, Appeals and Registers) Regulations, 1991 details the application procedures.

EXTENSION 18
The main prescribed processes are listed on page 85 of this workbook.

Some examples of the processes listed in the regulations are:

- petrochemical processes and acid manufacturing processes in the chemical industry;
- asbestos processes and fibre processes in the minerals industry;
- iron and steel processes and smelting processes in the minerals industry.

4.3 Prescribed substances

Normally, controlling a prescribed process will result in the control of any noxious substance emitted. However, there are also regulations specifically applying to the release of prescribed substances. Examples of these include:

Releases into the air:	Oxides of sulphur and other sulphur compounds. Metals, metalloids and their compounds. Asbestos and glass fibres.
Releases into water:	Mercury and its compounds. Cadmium and its compounds. Polychlorinated biphenyls (PCBs).
Releases to land:	Organic solvents. Phosphorus. Pesticides.

As you will discover when we discuss water pollution, there are other kinds of pollution, which are dealt with separately.

4.4 BATNEEC

Under integrated pollution control, the principle of **BATNEEC – best available technique not entailing excessive cost** is employed.

What does this mean? Suppose a manufacturer is carrying out one of the prescribed processes. It may be possible to reduce the pollution from the process by employing better techniques or equipment. It would then be up to the organization to prove that the costs of these new techniques or equipment would outweigh the benefits they would bring. Even if the Environment Agency agree, the organization may still be required to phase in the new equipment over a period of time.

No prescribed process can be carried out and no substance emitted without authorization from the appropriate environment agency. Anyone carrying out a prescribed process without authorization **will be guilty of a criminal offence**.

5 How the law is enforced

There are a number of official bodies involved in environmental protection. We will take a brief look at some of the most important ones and discuss their different roles and functions.

5.1 The Environment Agency for England and Wales, and the Scottish Environment Protection Agency

These Environment Agencies, as we have already seen, are tasked with implementing an integrated and coherent approach to pollution control and have overall responsibility for doing this. The Agency has responsibility for the regulation of pollution for most hazardous activities, including:

- operating the system of integrated pollution control (IPC);
- control of radioactive substances;
- monitoring waste disposal;
- control of air pollution;
- control of water pollution.

5.2 Sewage undertakers

The licensing body for controlling discharges to sewers is the privatized sewage undertaker, which is able to grant permission for industrial discharges, called 'trade effluent consents'. This is an unusual case of a private organization taking on the role of regulation of pollution. Appeals go to the Director General of Water Services.

5.3 Local authorities

Local authorities have responsibilities for a number of environmental matters:

- Town and country planning

  The local authority is normally the local planning authority. It also has responsibility for conservation areas, listed building protection and protection of the countryside.

- Waste disposal

  County councils are the waste disposal authorities, and are the bodies which deal with waste disposal applications.

■ Public health

Local authorities have wide powers under the Public Health Acts, including control of nuisances, noise and litter.

■ Air pollution

Local authorities have responsibility for control of smoke, fumes, dust etc.

5.4 Prosecution and punishment

Traditionally, the 'British approach' to enforcement of environmental laws has been one of 'co-operation rather than confrontation'. This attitude stems from the belief that the main aim of pollution control is to prevent harm to the environment or harm to humans, rather than to detect and punish.

Until fairly recently, the regulatory bodies were very reluctant to use the ultimate sanction of prosecution and punishment under the law. Most pollution offences are the result of accidents, and the enforcers have placed more emphasis on the intention of the polluters, rather than the offences themselves. Only when companies have seemed to deliberately and persistently flout the law have stronger measures been taken.

However, the EU has placed regulatory bodies under increasing pressure to prosecute offenders. Public opinion also supports this approach. The Environment Agency, and its immediate predecessors HMIP (Her Majesty's Inspectorate of Pollution) and the NRA (National Rivers Authority), have shown that they intend taking a stronger line. For example, in the eighteen months after the privatization of the water companies in 1989, the NRA prosecuted five of the ten companies, and Shell (UK) was fined £1 million in the Crown Court for polluting the River Mersey.

Activity 27

What is the lesson here for work organizations?

_____

_____

It seems that

companies and other organizations in the future will have to take much more care in preventing pollution and protecting the environment if they are to avoid prosecution.

5.5 The powers of inspectors under IPC

Under the system of integrated pollution control, the Environment Agency for England and Wales and the Scottish Environment Protection Agency both have very wide powers. If you scan through the following list you will get an idea of just how much power the law has given to environmental inspectors, who may:

- enter premises; although this is usually done at a reasonable time, entry can be made at any time where there is a risk of serious pollution;
- examine and investigate any process contained in any premises;
- direct that any premises and items contained on those premises remain undisturbed;
- take measurements, photographs and make recordings;
- take samples of air, water, articles or substances on or in the vicinity of the premises;
- require any articles or substances to be dismantled or subjected to any process or test;
- take possession and detain any article or substance;
- require persons to answer questions if the inspector reasonably believes them to be able to give relevant information and, once they have supplied answers, to sign a declaration as to the truth of those answers;
- require the production of any records;
- require the provision of any substance or facilities necessary to carry out any of the duties mentioned above;
- seize and render harmless any article or substance which is believed to be the cause of imminent danger.

6 The law on waste management

Some 500,000,000 tonnes of waste are produced annually in Britain; this figure gives some indication of the size of the waste management problem. Waste comes from domestic premises, industry, farms, mines, quarries, sewage, power stations and other sources.

Activity 28

2 mins

Write down **one** difficulty associated with the disposal of waste.

_____

_____

EXTENSION 19
A Department of Environment Code of Practice on waste management is available, and is listed in the extension.

Most waste is disposed of in large holes in the ground – so-called landfill sites. The main problem with doing this is that many substances break down over a period of time and create hazardous substances, thus polluting the land and nearby water environments. Other methods of disposal include incineration and dumping at sea, both of which methods have their problems.

The Environmental Protection Act, 1990 introduced a new system of waste management, and placed a 'duty of care' on all businesses to prevent improper disposal of waste. Under the EPA:

'any business which produces, imports, stores, treats, processes, transports, recycles or disposes of controlled waste must, by law, take all reasonable steps to look after any waste it has and prevent its illegal disposal by others'.

Controlled waste is any household, commercial or industrial waste, such as waste from a house, shop, office, factory, building site or any other business premises.

When waste changes hands, a transfer note must be completed and signed by both parties and a written description of the waste handed over. (This does not apply to household waste.)

7 The law on water pollution

The water environment comprises rivers, streams, natural and artificial lakes, underground waters and coastal areas.

Activity 29

3 mins

How many sources of water pollution can you think of? For instance, one source is industrial discharge by factories. Try to list **two** others.

_____

_____

_____

_____

Besides industrial emissions, you might have mentioned:

■   Farms

Fertilisers or silage plants sometimes leach into nearby streams and rivers; this gives rise to a process called **eutrophication** in which the nitrates and phosphates in these materials cause an excessive growth of algae. Pesticides and herbicides may also find their way into water courses from farms.

- Sewage works

  Many sewage works have consent to discharge their contents directly into coastal waters and other water environments.

- Waste sites

  Natural rain can cause leaching of noxious materials from waste tips.

- Mines

  The waste waters from mines may contaminate water courses.

- Accidents

  If chemicals or other substances are being transported and an accident occurs, they may flow directly into rivers or lakes.

- People

  It isn't unknown for people to throw away unwanted rubbish into rivers, streams and canals.

7.1 The system of 'consents'

If any organization wants to discharge any trade or sewage effluent into any inland or coastal waters, it can only do so when given a consent by the Environment Agency for England and Wales, or the Scottish Environment Protection Agency. Trade effluents includes discharge from farms, fish farms or from industrial plants. A consent also has to be granted before any discharge through pipes into the sea.

It is an offence to 'cause or knowingly permit' any such discharge unless it is carried out with a consent.

The 'polluter pays' principle applies and all costs by the enforcing agency will be recovered from the organization making the discharge.

7.2 The new water companies

Under the Water Act, 1989 the regional water authorities were privatized. Ten water services companies were set up, having responsibility for water supply and sewage services. Twenty-nine other smaller ones are responsible for water supply only.

Although the Water Act, 1989 set up the new structure for the industry, it was replaced in December 1991 by five new acts. These made hardly any change to the law, however. The new Acts are the:

- Water Industry Act, 1991;
- Water Resources Act, 1991;
- Water Companies Act, 1991;
- Land Drainage Act, 1991;
- Water Consolidation (Consequential and Amendments) Act, 1991.

8 The law on atmospheric pollution

Air is essential for humans and most other land creatures. Pollution of the air can have devastating consequences.

Activity 30

3 mins

Which industries would you expect to be the worst air polluters?

_____

_____

_____

List **two** or **three** substances that are pumped into the atmosphere by industry, and which are considered as being detrimental to the environment.

_____

_____

_____

Many types of industry use processes that discharge undesirable substances. The power generation industry is one that is commonly focused upon, because it is so large, and burns so much fuel.

For example, over a period of five years, the average thermal efficiency of coal-burning stations of one national company was raised from 35.37 per cent to 36.10 per cent. This does not seem to be a great improvement, until you realize that the saving reduced coal consumption by 800,000 tonnes, and cut carbon dioxide emissions by over 2 million tonnes.

Apart from carbon dioxide, which contributes to global warming, other substances emitted by power stations and factories burning coal and oil include:

- carbon monoxide;
- sulphur dioxide;
- nitric oxide;
- nitrogen dioxide;
- nitrous oxide;
- hydrochloric acid;
- particulate material;
- mercury.

Problems of air pollution have been written about since at least the seventeenth century, but with the coming of the industrial revolution the problems grew worse. The universal adoption of coal as a fuel meant that the smoking chimney became the symbol of industry. Acid emissions from caustic soda factories caused the first acid rain.

The effects on the countryside surrounding industrial towns were desolating: often trees just could not survive. Smog was commonplace from Victorian times. The effects of breathing this polluted atmosphere can be imagined.

8.1 Control of smoke

The Clean Air Acts of 1956 and 1968 provided a control mechanism for smoke, dust and fumes. These Acts prohibited the emission of 'dark smoke' from the chimney of any building, whether domestic or industrial; offenders were guilty of a criminal offence.

The height of chimneys was also brought under control in the 1968 Act. The idea was to increase the height of chimneys so as to disperse emissions over a wider area.

Activity 31

3 mins

What effects do you think that simply increasing the height of chimneys had on the environment?

_____

_____

_____

Increasing chimney heights, unless something is done about the problem of the emissions themselves, certainly spreads the pollution over a wider area. This may alleviate the local problem but, as is now recognized, passes on the pollution to another area or another country.

8.2 Air pollution control

The Environmental Protection Act, 1990 introduced a system of air pollution control (APC).

APC is a similar system to IPC (integrated pollution control) but covers only the less polluting substances, and is controlled by local authorities rather than the environment agencies. It includes:

- lower grade combustion processes;
- small iron and steel furnaces;
- low grade waste incineration;
- animal and vegetable treatment processes.

As with IPC, anyone carrying out a prescribed process or emitting a prescribed substance which is subject to air pollution control must have an authorization to do so. It is a criminal offence to break this law.

8.3 The control of vehicle emissions

The emission of pollutants from road vehicles is controlled by the Road Traffic Act, 1988 together with regulations relating to the construction of vehicles and type approval.

The regulations have been changed regularly to take account of EU directives, such as those on carbon monoxide emissions.

All cars manufactured after 1 October 1989 have had to be capable of running on unleaded petrol.

All cars with an engine size over 1000 cc manufactured from 1992 onwards have had to be fitted with a three-way catalytic converter. This device is fitted to the exhaust of the vehicle, and considerably reduces polluting emissions. Catalytic converters can only be used with unleaded petrol.

Self-assessment 3

20 mins.

1 Explain briefly what is meant by the principle of 'the polluter pays'.

_____

_____

_____

2 What are the **four** principal areas of environmental law?

_____

_____

_____

_____

3 Fill in the blanks in the following statements with suitable words.

The main objectives of IPC are:

■ to prevent or minimize the release of _____ _____

and to render _____ any such substances which are released;

■ to develop an approach to _____ control that considers

discharges from _____ processes to all media in the context of

the _____ as a whole.

4 Explain briefly what BATNEEC (best available technique not entailing excessive cost) means.

_____

_____

_____

5 Which of the following statements are correct? Tick the correct ones.

a The Environment Agency for England and Wales, or the Scottish Environment Protection Agency, have the right to bring an organization's work to a complete halt. ☐

b One of the key policies of new legislation is the lowering of chimney heights. ☐

c Among the list of industrial air pollutants are nitric oxide; nitrous oxide; and hydrochloric acid. ☐

d Under no circumstances would sewage works obtain consent to discharge their contents directly into water environments. ☐

e One problem with dumping waste on landfill sites is that many substances break down over a period of time and create hazardous substances, thus polluting land and water environments. ☐

f It is acceptable for organizations to pollute the environment, provided they pay for clearing up the mess afterwards. ☐

Answers to these questions can be found on pages 88–9.

9 Summary

- Environmental law has undergone some radical changes in the last twenty years or so.

- Like health and safety, environmental law typically consists of framework acts, which give rise to regulations that spell out the detail.

- The environmental policies of the EU have already greatly influenced British environmental legislation, and are likely to continue to do so in the future. The basic principles of these policies are that:
  - preventative action is to be preferred to remedial measures;
  - environmental damage should be rectified at source;
  - the polluter should pay for the costs of the measures taken to protect the environment;
  - environmental policies should form a component of the EU's other policies.

- The four principal areas of environmental law are:
  - pollution;
  - water;
  - nature conservation;
  - town and country planning.

- Recent legislation includes the Environmental Protection Act, 1990 (EPA), and the Environment Act, 1995.

- The main objectives of integrated pollution control (IPC) are:
  - to prevent or minimize the release of prescribed substances and to render harmless any such substances which are released;
  - to develop an approach to pollution control that considers discharges from industrial processes to all media in the context of the environment as a whole.

- Under integrated pollution control, the principle of BATNEEC – best available technique not entailing excessive cost – is employed.

- The main enforcement agency is the Environment Agency for England and Wales. In Scotland it is the Scottish Environment Protection Agency. Both have very wide powers, and are under pressure from the EU to prosecute offenders.

- The Environmental Protection Act, 1990 introduced a new system of waste management, and placed a 'duty of care' on all businesses to prevent improper disposal of waste.

- If any organization wants to discharge any trade or sewage effluent into any inland or coastal waters, it can only do so when given a consent by one of the two environment agencies.

- The Environmental Protection Act, 1990 introduced a system of air pollution control (APC). APC is a similar system to IPC (integrated pollution control) but covers only the less polluting substances, and is controlled by local authorities rather than the environment agencies.

Performance checks

1 Quick quiz

Jot down the answers to the following questions on *Managing Lawfully (Health, Safety and Environment)*

Question 1 What is the main difference between criminal law and civil law, in regard to the standard of proof required to decide a case?

_____

_____

_____

Question 2 What are the **three** routes by which an organization might have a legal action brought against it, as a result of an accident at work?

_____

_____

_____

Question 3 When a directive is issued by the EU, what actions are taken by member states?

_____

_____

Question 4 One thing an employer must do, in order to comply with HSWA, is to ensure plant and equipment are safely installed, operated and maintained. Give **two** other examples of what it must do.

_____

_____

Question 5 How would you summarize the responsibilities of employees under HSWA?

_____

_____

Question 6 If a regulation uses the words 'the employer shall', how should that be interpreted?

_____

_____

75

Question 7 How would you explain what it means to carry out a risk assessment, in a sentence or two?

_____

_____

Question 8 Briefly, what is the purpose of health surveillance?

_____

_____

Question 9 How would you define a 'competent person' who is to help an employer comply with health and safety laws?

_____

_____

Question 10 Which **two** sets of regulations are important in respect of workstations?

_____

_____

Question 11 What's the first thing you should consider, if you plan to move a heavy load manually?

_____

_____

Question 12 List **two** requirements for employers under the Personal Protective Equipment at Work Regulations 1992 (PPEWR).

_____

_____

Question 13 How would you define a hazardous substance, under COSHH 2?

_____

_____

Question 14 Write down **one** of the four EU basic policy principles, in respect of environmental legislation.

_____

_____

Question 15 What must an organization do before it discharges any trade or sewage effluent into inland or coastal waters?

_____

_____

Answers to these questions can be found on pages 89–90.

2 Workbook assessment

Read the following case incident and then deal with the instruction that follows, writing your answers on a separate sheet of paper.

■ Carthorne Computers was started by Mike Carthorne four years ago, and is still a small organization, with some eighteen employees. The business was originally housed in Mike's garage, and now has more spacious premises on an industrial estate. Mike and Karl Jacken, who went to school together, are interested primarily in the technicalities of the work, and together they designed the firm's main product – a device that networks computers without wiring.

Karl's wife looks after the accounts, and the other employees are mainly operators and technicians who build the devices. Mike does most of the marketing himself. While the company is small, all profits are being ploughed back into the business, and money is always tight.

The work is not dangerous, and there had been no accidents until one day when a woman employed as a packer tripped over a trailing lead, and had to have stitches in a head wound.

This incident got Mike and Karl thinking about health and safety law. They had to admit that they knew little about the law, and were rather worried. Could they get prosecuted over this accident, or any other mishap on the premises? If so, what might be the likely outcome? How could they ensure that they were acting within the law in their work activities?

Imagine you have been called in as a consultant. First, explain to Mike and Karl whether, and how, they might be prosecuted.

Then write down a list of ten to fifteen questions that you would like answers to, in regard to the activities of the company and the way it is organized for health and safety. These questions are intended to help an expert like you to give the company the best general advice. An example of a question might be: 'Do your employees use any hazardous substances in their work?' (You are not expected to go on to provide the advice – only to frame suitable questions. Also, don't concern yourself with the details of the packer's accident.)

Portfolio
of evidence
A1.2

3 Work-based assignment

60 mins

The time guide for this assignment gives you an approximate idea of how long it is likely to take you to write up your findings. You will find you need to spend some additional time gathering information, talking to colleagues, and thinking about the assignment.

Your written response to this assignment may form useful evidence for your S/NVQ portfolio. The assignment is designed to help you to demonstrate your Personal Competence in:

- building teams;
- focusing on results;
- thinking and taking decisions;
- striving for excellence.

What you have to do

Select one of the regulations we discussed in the workbook, and carry out a brief investigation into how well your part of the organization complies with it. Start by drawing up a checklist. Then, by talking with your team and other colleagues, and by making observations, decide whether (and, if appropriate, how well) the law is being complied with.

Once you have done that, decide what steps need to be taken in order to ensure that any areas that are falling short of full compliance (or failing altogether) can be brought up to scratch.

Write up your findings and recommendations in the form of a report to your manager.

Reflect and review

Now that you have completed your work on *Managing Lawfully – Health, Safety and Environment*, let us review our workbook objectives.

The first objective was:

■ When you have completed this workbook you will be better able to identify the most important laws related to health and safety.

We have covered a lot of ground in this workbook: no fewer than sixteen different statutes related to health and safety were reviewed. These are the laws that are currently the most important to work organizations generally.

■ What can you do to increase your knowledge of health and safety law?

_____

_____

The second objective was:

■ When you have completed this workbook you will be better able to find out more about laws that are especially relevant to the work you do.

If you followed the workbook carefully, you should now have at least some idea about the kind of laws that deal with your type of work and organization. Now's the time to follow these up.

■ Write down the laws that you know about, that are particularly relevant to your work.

_____

_____

_____

_____

_____

_____

_____

_____

79

■ Now write down the actions you plan to take to find out more about these and (perhaps) other laws.

_____

_____

_____

_____

_____

_____

_____

_____

The third objective was:

■ When you have completed this workbook you will be better able to explain to your team how the law affects them, and the duties imposed by the law on everyone at work.

A good starting point would be the Health and Safety at Work etc. Act, 1974, which makes specific reference to employees' duties. Then you should consider the other 'six-pack' regulations, and COSHH 2 if relevant, all of which we covered in Session B.

■ Are you clear in your own mind about these laws, at least so far as to be able to pass on the relevant information?

_____

_____

■ If you are still confused about the law, what steps do you intend to take to clarify your thinking on the subject?

_____

_____

■ If your team is in need of further information and training on health and safety, what plans will you make for them to receive it?

_____

_____

The final objective was:

■ When you have completed this workbook you will be better able to understand the law on the environment.

Session C was concerned with environmental law. We took a brief look at the four aspects of the environment: water, pollution; nature conservation; and town and country planning. In addition, we discussed the two main laws: the

Environmental Protection Act, 1990, and the Environment Act, 1995, and mentioned a number of others.

Like health and safety law, environmental law imposes a great number of requirements on organizations and employers. To a large extent, the law reflects public concern over these issues. So, again in common with health and safety, it is not usually enough simply to comply with the law. The public image of an organization can deteriorate dramatically if it is not seen to be taking a constructive attitude, and taking positive actions.

■ Be honest: how well do you understand environmental law, at least so far as it affects your activities at work?

_____

_____

■ What steps do you intend to take to increase your knowledge and awareness of environmental issues?

_____

_____

2 Action plan

Use this plan to further develop for yourself a course of action you want to take. Make a note in the left-hand column of the issues or problems you want to tackle, and then decide what you intend to do, and make a note in Column 2.

The resources you need might include time, materials, information or money. You may need to negotiate for some of them, but they could be something easily acquired, like half an hour of somebody's time, or a chapter of a book. Put whatever you need in Column 3. No plan means anything without a timescale, so put a realistic target completion date in Column 4.

Finally, describe the outcome you want to achieve as a result of this plan, whether it is for your own benefit or advancement, or a more efficient way of doing things.

Desired outcomes

1 Issues	2 Action	3 Resources	4 Target completion

Actual outcomes

3 Extensions

Extension 1   Book   *Health and Safety Law*
                    Author   Jeremy Stranks
                    Edition   2nd edition 1996
                    Publisher Pitman

Extension 2   Book   *A Guide to the Health and Safety at Work etc. Act, 1974*
                    Edition   1992
                    Publisher HSE Books

Extension 3   Book   *Successful Health and Safety Management*
                    Edition   1991
                    Publisher HSE Books

Extension 4   Book   *Safety Representatives and Safety Committees (The Brown Book)*
                    Edition   1996
                    Publisher Health and Safety Executive

Extension 5   Book   *Management of Health and Safety at Work Regulations, 1992 (Approved Code of Practice)*
                    Edition   1992
                    Publisher Health and Safety Commission

Extension 6   Book   *Workplace Health, Safety and Welfare (Approved Code of Practice)*
                    Edition   1992
                    Publisher Health and Safety Commission

Extension 7   Book   *Manual Handling – Guidance on Regulations. Manual Handling Operations Regulations, 1992*
                    Edition   1992
                    Publisher Health and Safety Executive

Extension 8   Book   *Display Screen Equipment Work – Guidance on Regulations. Health and Safety (Display Screen Equipment) Regulations, 1992*
                    Edition   1992
                    Publisher Health and Safety Executive

Extension 9	Book	*Personal Protective Equipment at Work – Guidance on Regulations. Personal Protective Equipment at Work Regulations, 1992*
	Edition	1992
	Publisher	Health and Safety Executive
Extension 10	Book	*Work Equipment – Guidance on Regulations. Provision and Use of Work Equipment Regulations, 1992*
	Edition	1992
	Publisher	Health and Safety Executive
Extension 11	Book	*General COSHH (Approved Code of Practice)*
	Edition	1995
	Publisher	Health and Safety Executive

Extension 12   For mail order of books from the Health and Safety Executive, write to: HSE Books, PO Box 1999, Sudbury, Suffolk, CO10 6FS. Tel: 01787 881165. Fax: 01787 313995.

For information, call the HSE InfoLine. Tel: 0541 545500. Or write to: HSE Information Centre, Broad Lane, Sheffield S3 7HQ.

Extension 13	Book	*Memorandum of Guidance on the Electricity at Work Regulations, 1984, Revised*
	Edition	1989
	Publisher	Health and Safety Executive
Extension 14	Book	*Electricity at Work: Safe Working Practices*
	Edition	1993
	Publisher	Health and Safety Executive
Extension 15	Book	*First Aid at Work. Health and Safety (First-Aid) Regulations, 1981 (Approved Code of Practice)*
	Edition	1990
	Publisher	Health and Safety Executive
Extension 16	Book	*A Guide to RIDDOR 95*
	Edition	1996
	Publisher	Health and Safety Executive
Extension 17	Book	*A Guide to RIDDOR 95 Plus Electronic Versions of Forms F2508/F2508A on Diskette*
	Edition	1996
	Publisher	Health and Safety Executive

Extension 18   List of prescribed processes affected by integrated pollution control.

Mineral Industry

Cement
Asbestos
Fibre
Glass
Ceramic

Fuel and Power Industry

Combustion (>50MWth):
boilers and furnaces
Gasification
Carbonization
Combustion (remainder)
Petroleum

Chemical Industry

Petrochemical
Organic
Chemical pesticide
Pharmaceutical
Acid manufacturing
Halogen
Chemical fertilizer
Bulk chemical storage
Inorganic material

Waste Disposal Industry

Incineration
Chemical recovery
Waste-derived fuel

Other Industry

Paper manufacturing
Di-isocyanate
Tar and bitumen
Uranium
Coating
Coating manufacturing
Timber
Animal and plant treatment

Metal Industry

Iron and steel
Smelting
Non-ferrous

Extension 19   Book   *Waste Management – The Duty of Care (A Code of Practice)*
Publisher   HMSO

Many of these Extensions can be taken up via your NEBS Management Centre. They will either have them or will arrange that you have access to them. However, it may be more convenient to check out the materials with your personnel or training people at work – they may well give you access. There are other good reasons for approaching your own people; for example, they will become aware of your interest and you can involve them in your development.

4 Answers to self-assessment questions

**Self-assessment 1
on page 23**

1 The correct statements are:

b Contract law is relatively unimportant in health and safety matters.

c Following an accident, an organization may be prosecuted under either criminal law or civil law.

d European law takes precedence over UK law.

f The Health and Safety at Work etc. Act, places an obligation on employers to take care of the health and safety of customers on its premises.

g Employees have duties to co-operate with employers in meeting the requirements of the law.

i 'As far as reasonably practicable' means that the degree of risk can be balanced against the cost of taking measures to avoid the risk.

2 Safety representatives may be involved in:

■ TALKING to employees about particular health and safety problems;

■ encouraging CO-OPERATION between their employer and employees;

■ carrying out INSPECTIONS of the workplace to see whether there are any real or potential HAZARDS that haven't been adequately addressed;

■ bringing to the attention of the employer any UNSAFE or unhealthy conditions or working PRACTICES, or unsatisfactory WELFARE arrangements;

■ REPORTING to employers about these problems and other matters connected to health and safety in that WORKPLACE;

■ taking part in ACCIDENT investigations.

3 a Approved codes of practice (ACOPs) – [ix] Issued by the Health and Safety Commission (HSC) as interpretations of regulations, and are intended to help people apply the law in practice.

b Civil law – [iv] A plaintiff sues a defendant, usually for damages, that is, financial compensation. As an example, an individual may sue an employer if he or she is injured at work.

c Common law – [ii] Based on case law: the decisions made in courts over the centuries. Once a judgement is made, a precedent is established.

d County courts – [vii] They deal only with civil matters. More complex civil cases, or ones involving large sums of money, go instead to one of the three divisions of the High Court of Justice.

e Criminal law – [iii] Anyone committing a crime has offended against the state, and is in breach of this. If an organization fails to comply with its statutory health and safety duties, its officers may be prosecuted.

f Crown Courts – [vi] More serious cases go to one of these, and these courts also hear appeals from other courts.

g EU Directives – [viii] Bind member countries to comply with an agreed ruling. They are normally made into national laws by each state.

h Magistrates' courts – [v] Deal mainly with the less serious criminal offences.

i Statute law – [i] Acts of Parliament (such as the Health and Safety at Work etc. Act, 1974), together with a great many 'statutory instruments' or 'subordinate legislation'.

Self-assessment 2 on page 51

1

Under MHSWR, employers must:	This includes the process of:
a provide risk assessment;	iii identifying the hazard; measuring and evaluating the risk from this hazard; putting measures into place that will either eliminate the hazard, or control it.
b provide health surveillance;	i identifying adverse affects; rectifying inadequacies in control; informing those at risk of any damage to their health; reinforcing health education.
c appoint competent persons;	iv identifying those with sufficient training, and experience or knowledge and other qualities; requiring them to devise and apply the measures needed to comply with health and safety laws.
d consult employees' safety representatives.	ii identifying measures that may substantially affect health and safety; identifying health and safety aspects of new technology; discussing these with the relevant people.

2 More than one regulation will apply in most work situations. Some you may have identified are as follows.

i Loading bags of flour: [c] Manual Handling Operations Regulations, 1992 (MHOR).

ii Supervising telesales: [d] Health and Safety (Display Screen Equipment) Regulations, 1992.

iii Running an electrical department in a superstore: [b]Workplace (Health, Safety and Welfare) Regulations, 1992 (WHSWR).

iv Training fork-lift truck drivers in a fuel depot: [a] Management of Health and Safety at Work Regulations, 1992 (MHSWR).

87

    v   Supervising an area where there are high levels of dust. [a] Management of Health and Safety at Work Regulations, 1992 (MHSWR),

    vi Supervising on a building site: [e] Personal Protective Equipment at Work (PPE) Regulations, 1992 (PPEWR).

3   Under the COSHH 2 regulations, employers have to:

- determine the HAZARD of SUBSTANCES used by the organization;

- assess the RISK to people's health from the way the substances are used;

- prevent anyone being EXPOSED to the substances, if possible;

- if exposure cannot be prevented, decide how to CONTROL the exposure so as to reduce the risk, and then establish effective CONTROLS;

- ensure that the controls are properly used and MAINTAINED;

- examine and TEST the control measures, if this is required;

- inform, INSTRUCT and train employees (and non-employees on the premises), so that they are aware of the hazards and how to work SAFELY;

- if necessary, MONITOR the exposure of employees (and non-employees on the premises), and provide HEALTH SURVEILLANCE to employees if necessary.

Self-assessment 3 on page 72

1   The principle of the polluter pays means that producers of goods or services should be responsible for the costs of preventing or dealing with any pollution that they may cause.

2   The four principal areas of environmental law are:

- pollution;

- water;

- nature conservation;

- town and country planning.

3   The main objectives of IPC are:

- to prevent or minimize the release of **prescribed substances** and to render **harmless** any such substances which are released;

- to develop an approach to **pollution** control that considers discharges from **industrial** processes to all media in the context of the **environment** as a whole.

4   BATNEEC means that organizations must use the best available techniques in reducing or eliminating pollution, unless they can prove that the costs of these techniques would outweigh the benefits.

5   The correct statements are as follows.

a   The Environment Agency for England and Wales, or the Scottish Environment Protection Agency, have the right to bring an organization's work to a complete halt.

c   Among the list of industrial air pollutants are nitric oxide; nitrous oxide; and hydrochloric acid.

e   One problem with dumping waste on landfill sites is that many substances break down over a period of time and create hazardous substances, thus polluting land and water environments.

5   Answers to the quick quiz

Answer 1   A lesser standard of proof applies in civil actions: cases have to be proved 'on the balance of probabilities', rather than 'beyond reasonable doubt'.

Answer 2   The two main routes are through the civil and criminal courts. The third route is via an industrial tribunal.

Answer 3   Directives, which bind member countries to comply with an agreed ruling, are normally made into national laws by each state.

Answer 4   Other examples are: check systems of work frequently, to ensure that risks from hazards are minimized; monitor the work environment regularly, to ensure that people are protected from any toxic contaminants; inspect safety equipment regularly; minimize risks to health from 'natural and artificial substances'.

Answer 5   Employees have responsibilities under HSWA to take care for their own health and safety, and that of their colleagues; to co-operate in meeting the requirements of the law; not to interfere with or misuse anything provided to protect their health, safety and welfare.

Answer 6   If the words 'the employer shall' are used, it means that the requirement that follows is compulsory.

Answer 7   To carry out a risk assessment, you must identify the hazard; measure and evaluate the risk from this hazard; and put measures into place that will either eliminate the hazard, or control it.

Answer 8   The purpose of health surveillance is to identify adverse affects early; rectify inadequacies in control, and so reduce the risks to those affected or exposed; inform those at risk, as soon as possible, of any damage to their health, so that they can take action; to reinforce health education.

Answer 9   A person can be regarded as competent if he or she has 'sufficient training, and experience or knowledge and other qualities, properly to undertake' the role.

Answer 10   The Workplace (Health, Safety and Welfare) Regulations, 1992 (WHSWR), and the Health and Safety (Display Screen Equipment) Regulations, 1992.

Answer 11   You should consider whether the load must be moved at all, and if so, whether it could be moved by non-manual methods.

Answer 12   Under PPEWR, employers have to ensure this equipment is suitable and appropriate; maintain, clean and replace it; provide storage for it when not in use; ensure that it is properly used; give employees training, information and instruction in its use.

Answer 13   A hazardous substance is virtually any substance that can affect health.

Answer 14   The basic principles of EU environmental policies are that: preventative action is to be preferred to remedial measures; environmental damage should be rectified at source; the polluter should pay for the costs of the measures taken to protect the environment; environmental policies should form a component of the EU's other policies.

Answer 15   The organization must obtain a consent.

6 Certificate

Completion of this certificate by an authorized person shows that you have worked through all the parts of this workbook and satisfactorily completed the assessments. The certificate provides a record of what you have done that may be used for exemptions or as evidence of prior learning against other nationally certificated qualifications.

Pergamon Open Learning and NEBS Management are always keen to refine and improve their products. One of the key sources of information to help this process are people who have just used the product. If you have any information or views, good or bad, please pass these on.

Managing Lawfully
– Health, Safety and Environment

....

has satisfactorily completed this workbook

Name of signatory ....

Position ....

Signature ....

Date ....

Official stamp

SUPER SERIES

SUPER SERIES 3

0-7506-3362-X	Full Set of Workbooks, User Guide and Support Guide

A. Managing Activities

0-7506-3295-X	1. Planning and Controlling Work
0-7506-3296-8	2. Understanding Quality
0-7506-3297-6	3. Achieving Quality
0-7506-3298-4	4. Caring for the Customer
0-7506-3299-2	5. Marketing and Selling
0-7506-3300-X	6. Managing a Safe Environment
0-7506-3301-8	7. Managing Lawfully - Health, Safety and Environment
0-7506-37064	8. Preventing Accidents
0-7506-3302-6	9. Leading Change
0-7506-4091-X	10. Auditing Quality

B. Managing Resources

0-7506-3303-4	1. Controlling Physical Resources
0-7506-3304-2	2. Improving Efficiency
0-7506-3305-0	3. Understanding Finance
0-7506-3306-9	4. Working with Budgets
0-7506-3307-7	5. Controlling Costs
0-7506-3308-5	6. Making a Financial Case
0-7506-4092-8	7. Managing Energy Efficiency

C. Managing People

0-7506-3309-3	1. How Organisations Work
0-7506-3310-7	2. Managing with Authority
0-7506-3311-5	3. Leading Your Team
0-7506-3312-3	4. Delegating Effectively
0-7506-3313-1	5. Working in Teams
0-7506-3314-X	6. Motivating People
0-7506-3315-8	7. Securing the Right People
0-7506-3316-6	8. Appraising Performance
0-7506-3317-4	9. Planning Training and Development
0-75063318-2	10. Delivering Training
0-7506-3320-4	11. Managing Lawfully - People and Employment
0-7506-3321-2	12. Commitment to Equality
0-7506-3322-0	13. Becoming More Effective
0-7506-3323-9	14. Managing Tough Times
0-7506-3324-7	15. Managing Time

D. Managing Information

0-7506-3325-5	1. Collecting Information
0-7506-3326-3	2. Storing and Retrieving Information
0-7506-3327-1	3. Information in Management
0-7506-3328-X	4. Communication in Management
0-7506-3329-8	5. Listening and Speaking
0-7506-3330-1	6. Communicating in Groups
0-7506-3331-X	7. Writing Effectively
0-7506-3332-8	8. Project and Report Writing
0-7506-3333-6	9. Making and Taking Decisions
0-7506-3334-4	10. Solving Problems

SUPER SERIES 3 USER GUIDE + SUPPORT GUIDE

0-7506-37056	1. User Guide
0-7506-37048	2. Support Guide

SUPER SERIES 3 CASSETTE TITLES

0-7506-3707-2	1. Complete Cassette Pack
0-7506-3711-0	2. Reaching Decisions
0-7506-3712-9	3. Making a Financial Case
0-7506-3710-2	4. Customers Count
0-7506-3709-9	5. Being the Best
0-7506-3708-0	6. Working Together

To Order - phone us direct for prices and availability details
(please quote ISBNs when ordering)
College orders: 01865 314333 • Account holders: 01865 314301
Individual purchases: 01865 314627 (please have credit card details ready)

We Need Your Views

We really need your views in order to make the Super Series 3 (SS3) an even better learning tool for you. Please take time out to complete and return this questionnaire to Trudi Righton, Pergamon Flexible Learning, Linacre House, Jordan Hill, Oxford, OX2 8DP.

Name :..

Address :..

....

Title of workbook :..

If applicable, please state which qualification you are studying for. If not, please describe what study you are undertaking, and with which organisation or college:

....

Please grade the following out of 10 (10 being extremely good, 0 being extremely poor):

Content   ....   Appropriateness to your position   ....

Readability   ....   Qualification coverage   ....

What did you particularly like about this workbook?

....
....
....

Are there any features you disliked about this workbook? Please identify them.

....
....
....

Are there any errors we have missed? If so, please state page number: ....

How are you using the material? For example, as an open learning course, as a reference resource, as a training resource etc.

....

How did you hear about Super Series 3?:

Word of mouth: ☐   Through my tutor/trainer: ☐   Mailshot: ☐

Other (please give details):..

....

Many thanks for your help in returning this form.